3rd edition

geog.1

workbook

\<anna king\>\<jack mayhew\>\<susan mayhew\>\<justin woolliscroft\>

Name:

Class:

 OXFORD

Contents

1.1 Hey, you over there!

This is about where you live – planet Earth!

1 The earth spins as it travels non-stop around the sun. Why do you not fall off?

_____

_____

2 Planet Earth is full of life. Fill in the gaps in the following passage using words from the box below.

Elephants are the _____ animals on earth. There are only around 600 000

of them left: they are an _____ species. New forms of _____ are being

found all of the time. This new life is found on the land and also in the _____.

oceans	endangered	largest	life	animal

Did you know, for every one of me, there are 11 166 of you humans out there!

3 Choose _one_ fact about planet Earth. Draw a diagram in the box below to show your chosen fact.

1.2 Our planet: always changing

This is about how natural forces and humans change our planet

_____ _____ _____ _____

_____ _____ _____ _____

_____ _____ _____ _____

_____ _____ _____ _____

1 a Each of the photos shows how planet Earth is changing. Under each photo write a caption saying why. Choose from the captions given below.

> Water in the river scrapes and shapes the land as it flows
>
> Villages, towns and cities use a lot of land
>
> Hot rock can change what the landscape looks like
>
> Factories use valuable resources and sometimes cause pollution

b When you have written your captions, underline the natural reasons in one colour, and the human causes in another.

2 Humans are destroying the Earth. Do you agree?

_____

_____

_____

_____

1.3 ◀ **Your place on the planet**

This is about where you would like to live.

1 Everybody can think of things that would make a place great to live in. These are
often called the *features* of a place. In the box below circle five features that would
be important in your ideal place.

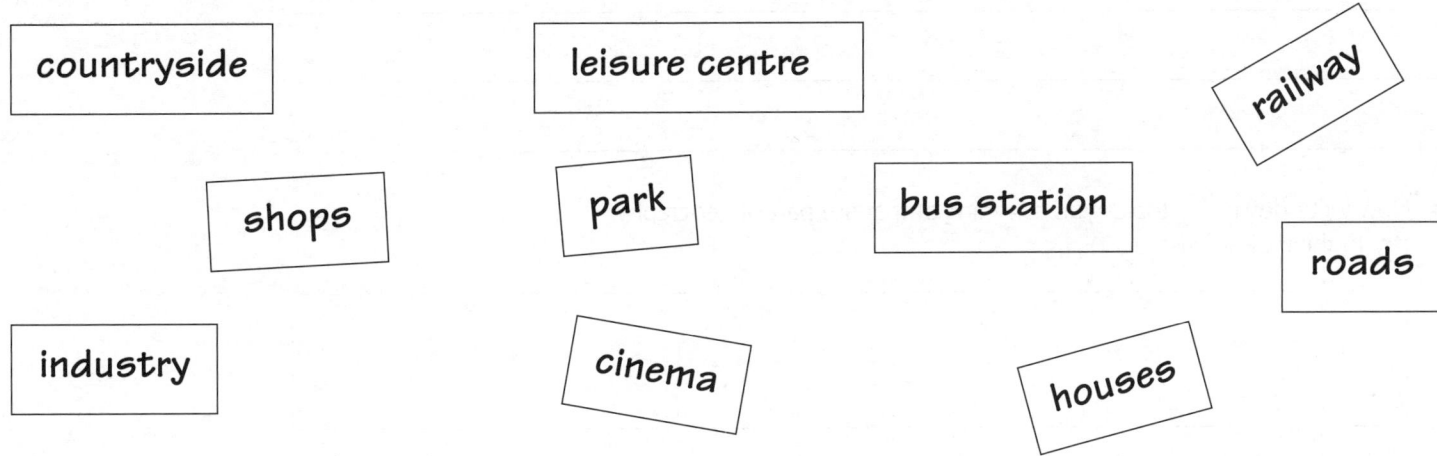

countryside

leisure centre

railway

shops

park

bus station

roads

industry

cinema

houses

2 In the space below make a sketch of your ideal place to live using the five features
that you have chosen.

3 Why have you planned your place like you have? What reasons can you think of to
explain? Give as many as you can.

_____

_____

_____

_____

1.4 ▸ It's all geography!

This is about how being nosy could make you a good geographer!

1   Geography can be divided into three different strands, *physical*, *human* and *environmental*. Explain what each one means in your own words.

_____

_____

_____

2 **a** Now write down any topics you can think of that are part of geography (try to think of at least eight).

[ ]

b Circle any topics that are *physical* geography in one colour, *human* geography in another, and environmental geography in a third.

c Do any of your topics include all three strands? Circle them in a fourth colour.

3 **a** Have a look at this photo. Underneath, brainstorm questions you could ask about it (can you think of six?).

_____

_____

_____

_____

_____

_____

_____

_____

b Are you able to answer any of your questions? How would you find answers to the others?

_____

_____

1 ◀ And to finish ...

Here are some of the things you'll study in geography.

a Colour all the boxes to do with *physical* geography in red.

b Colour all the boxes to do with *human* geography in blue.

c Colour all the boxes to do with *environmental* geography in green.

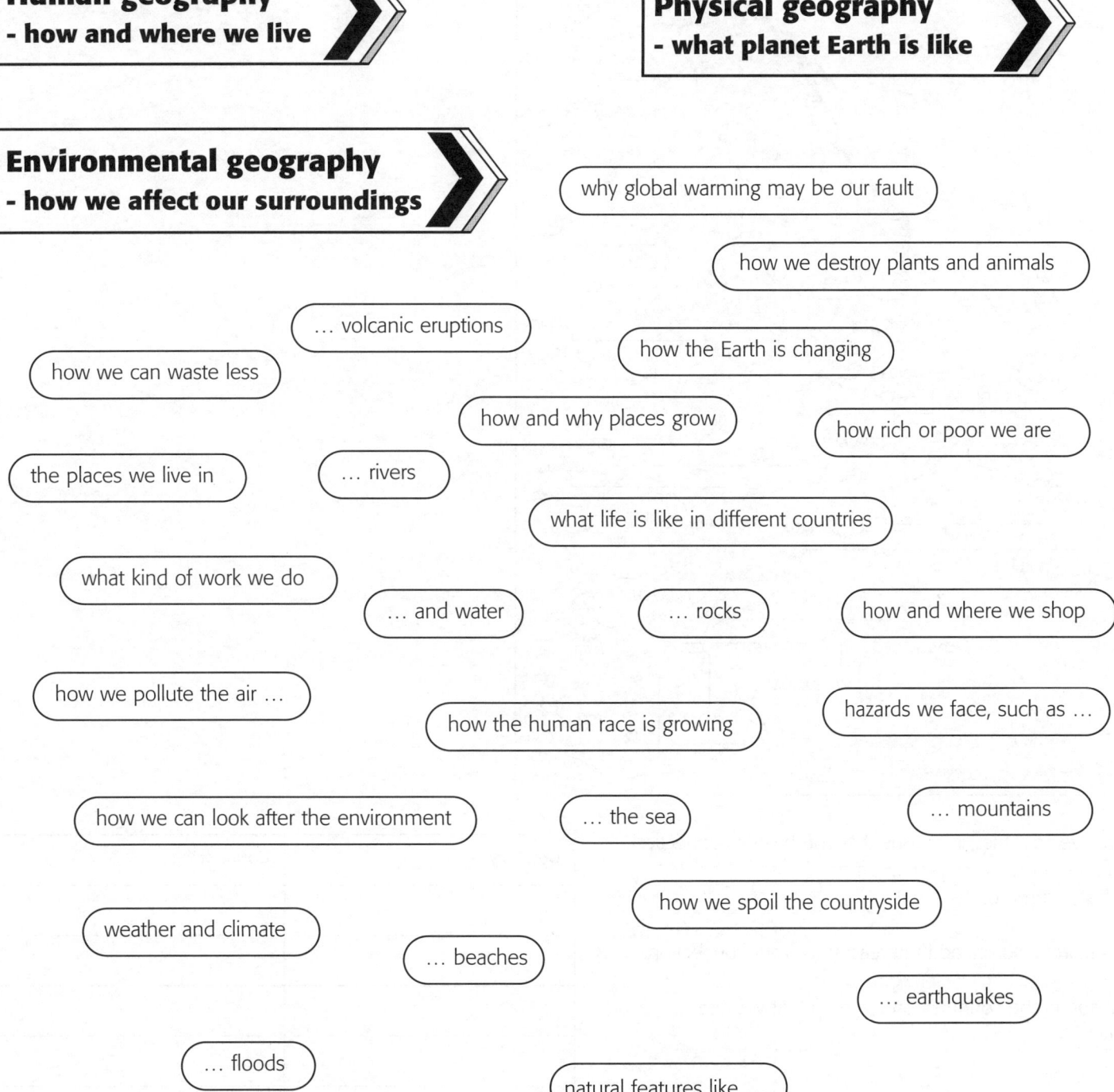

Human geography
- how and where we live

Physical geography
- what planet Earth is like

Environmental geography
- how we affect our surroundings

why global warming may be our fault

how we destroy plants and animals

... volcanic eruptions

how the Earth is changing

how we can waste less

how and why places grow

how rich or poor we are

the places we live in

... rivers

what life is like in different countries

what kind of work we do

... and water

... rocks

how and where we shop

how we pollute the air ...

how the human race is growing

hazards we face, such as ...

how we can look after the environment

... the sea

... mountains

how we spoil the countryside

weather and climate

... beaches

... earthquakes

... floods

natural features like ...

2.1 ◄ Making connections

This is about how we are connected to people and places all over the world – and how this can be shown using maps.

Like Walter, you are connected to hundreds of places. Some of them are in Europe.

a Think of five countries in Europe that you have a connection with. (Food, sport, holidays?). List these in the first column of the table.

b In the second column, add the reason for your connection.

c Now shade in the countries on the map that you are connected to.

Country	Connection

2.2 ◀ A plan of Walter's room

This is about what plans are, and what the scale of the plan tells you.

1 A drawing of something seen from above is called a **plan**.
Match the drawing of a chair with the correct plan (tick the correct one).

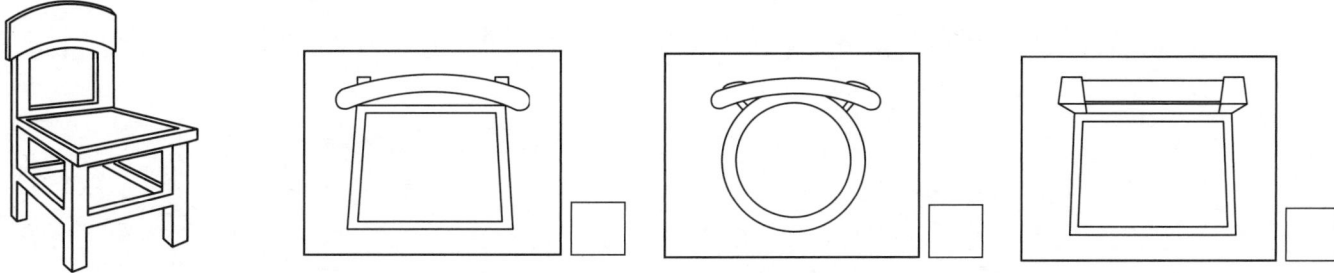

2 Now look at this plan of a bedroom. 1 cm on the plan represents 40 cm in the room. That is the **scale** of the plan.

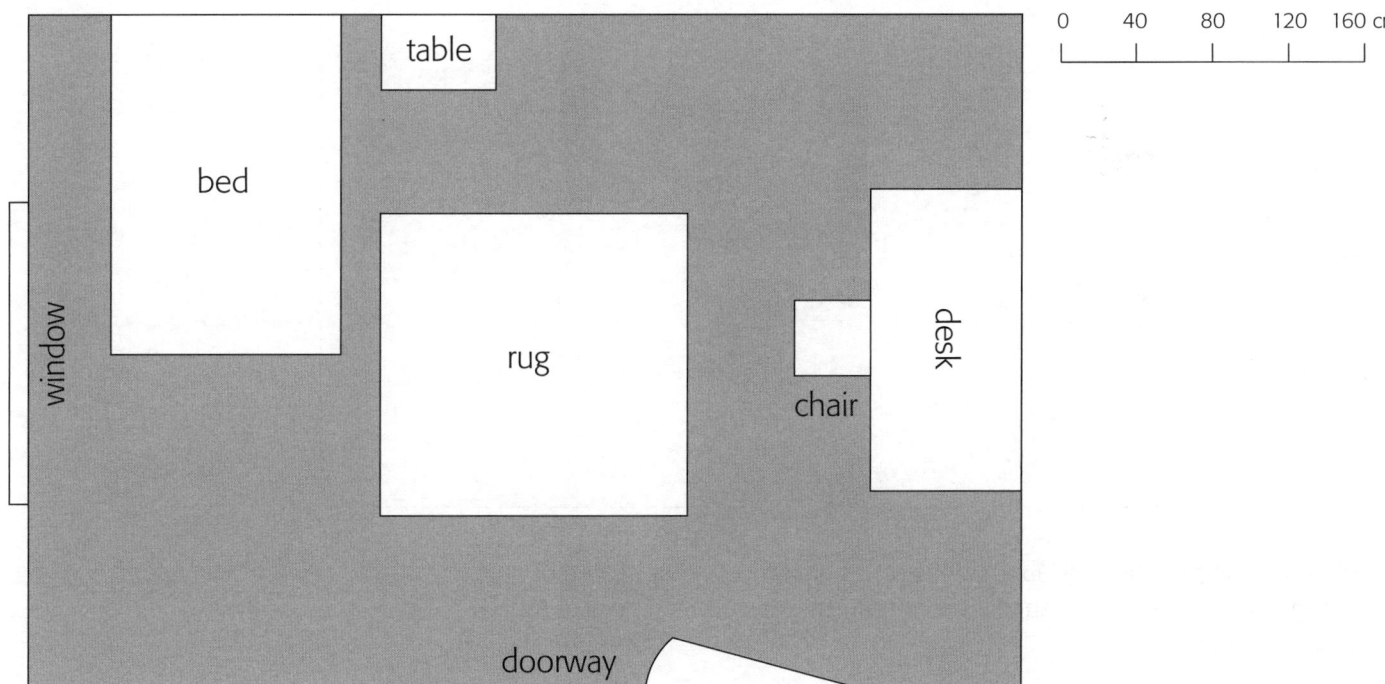

a On the plan, the window is this wide: _____

So, in real life the window is _____ cm wide. (Fill in the gap.)

b Now measure the length of the bed and fill in the gaps below.

On the plan, the bed is _____ cm long.

This means it is _____ cm long in real life.

c Something in the room is 60 cm wide in real life. What is it? _____

_____

2.3 ◀ **Your mental maps**

This is about your very own, personal mental maps.

1 a Think about a place where you like to go and play – it may be a park or a playground for example. In the space below draw a mental map of that area.

b Add labels to show your feelings about the various parts of your map. Some of your feelings may be happiness, anger, fear or sadness, but you may be able to think of others as well.

Think of a symbol for each of your feelings and draw the symbol in the correct place on your map.

c Show your map to a partner. Write down the thing that they like best about your map.

_____

_____

2.4 ‹ Real maps

This is about how maps are built up.

This photo shows a railway bridge over the River Tamar in Devon.
Your task is to finish this sketch map (which has been started for you) of the same place.
Don't forget a key!

Key

2.5 Using grid references

This is about finding places on a map, using grid references.

1 A good map has five things: a _Key_ , a frame around it, an arrow to show
Northing , a _Title_ and a _Scale_ .

title	note	north	scale
east	lock	key	

2 Look at this map.

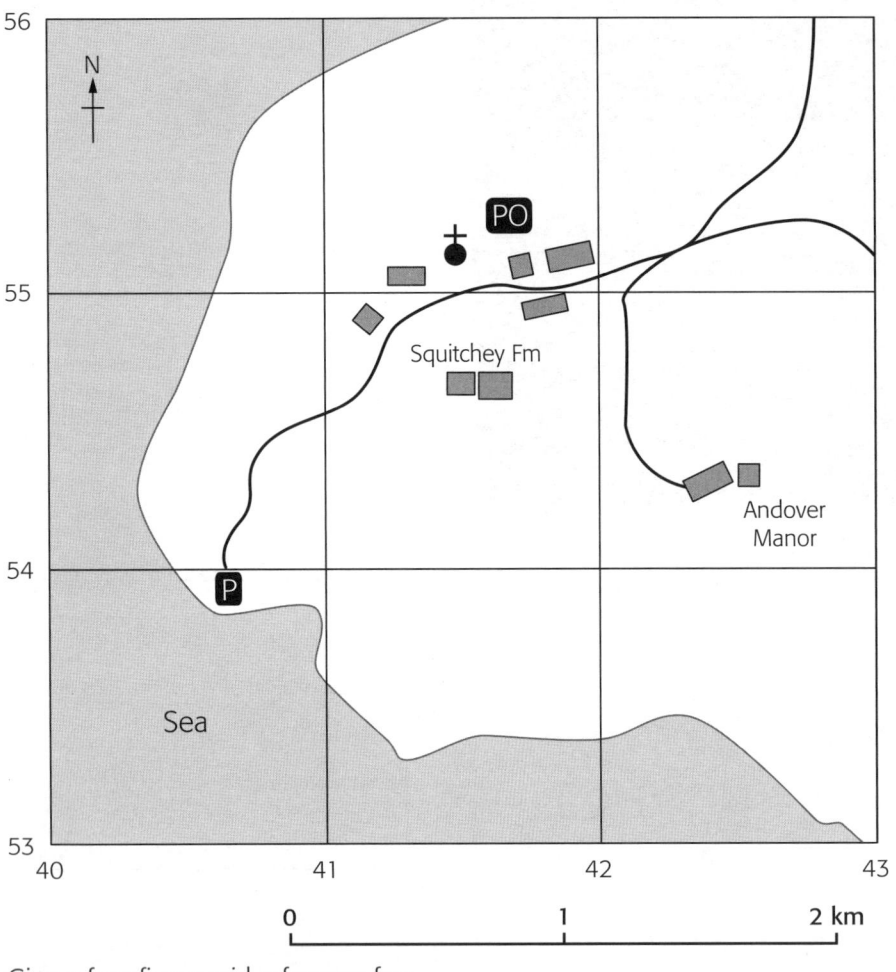

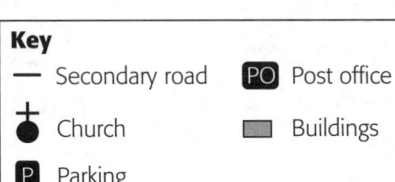

Give a four-figure grid reference for:

a Squitchey Farm _41,54_

b Andover Manor _42,54_

3 What is at this grid reference on the map?

a 407539 _Parking_ ~~~~

b 414552 _church_

4 How far is it from the church to Andover Manor by road? _1.5 Km_

2.6 ◄ How far?

This is about how to find the distance between two places on a map.

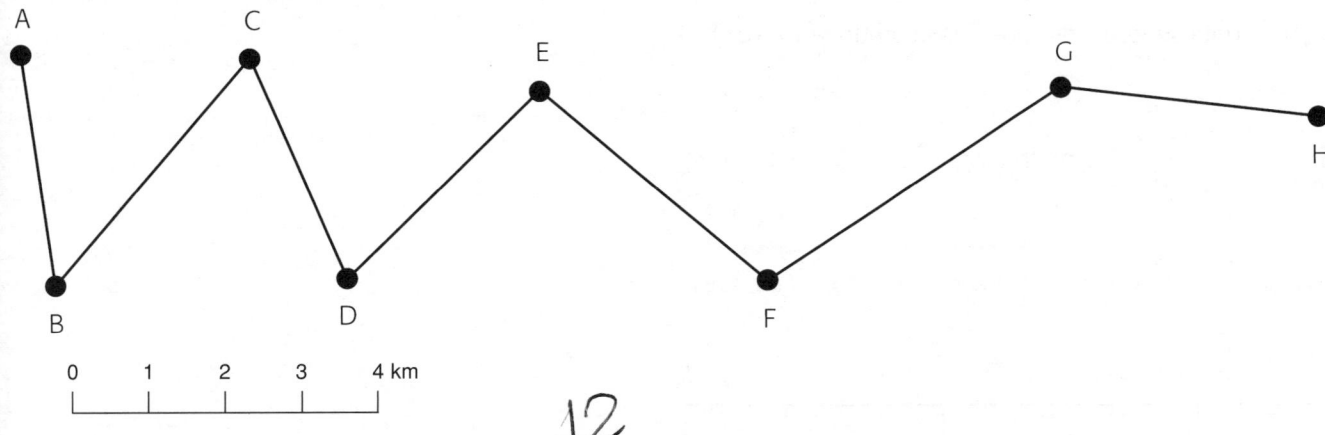

```
0    1    2    3    4 km
|____|____|____|____|
```

1 How far is it as the crow flies from A to H? _____ km

2 How far is it by road from A to H? _____

3 Have a look at the map on page 15.

   a How far is it as the crow flies from Hella Point (in square 3721) to Pordenack

   Point (square 3424)? _____

   b How far is it by road between St Buryan (4125) and Trethewey (3823)?

   _____

   c Follow these instructions.

   Drive east from Land's End for just over a kilometre. Take a right turning and follow the
   road for 1.8 km. Turn left and follow the short track to the end.

   Where do you end up? _____

   d Now give instructions (as in **c**) to someone who wants to travel from Treen (3922)
   to Trebehor (3724).

   _____

   _____

   _____

2.7 ◄ Which direction?

This is about how to give and follow directions, using N, S, E and W.

1 Draw an arrow going to these locations (one has been done for you).

○ ○ ◄─○

a to the south **b** to the north east **c** to the west

○ ○ ○

d to the south west **e** to the north west **f** to the east

2 Look at this grid.

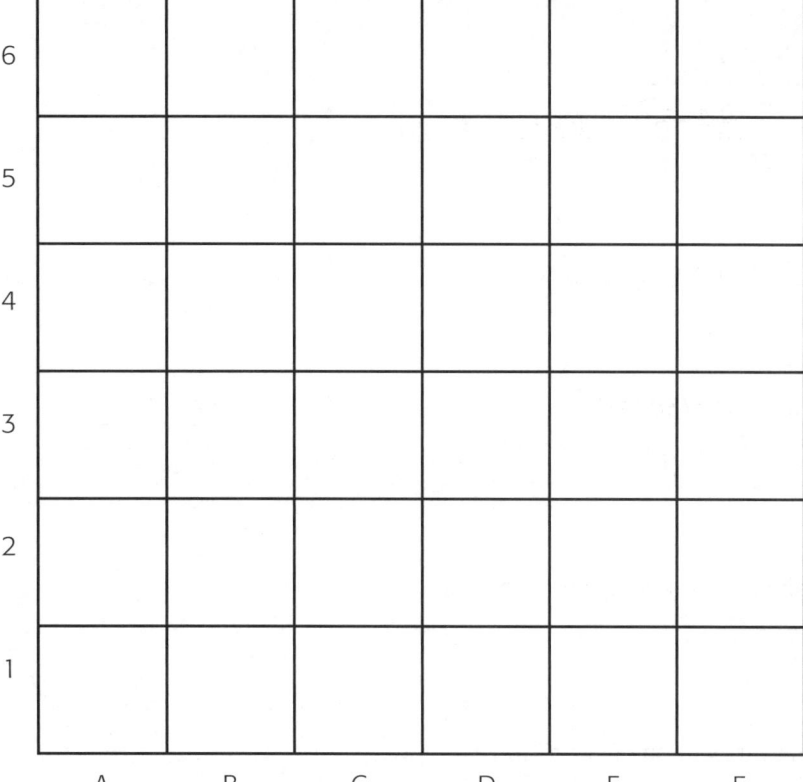

● Start in square E5.

● Go three squares W. Colour this square *red*.

● Go two squares S. Colour this square *green*.

● Go two squares SE. Colour this square *blue*.

● Go one square NE. Colour this square *yellow*.

● Go four squares NW. Colour this square *purple*.

● Go four squares E and one square S. What square do you end up in?

_____

2.8 ► Ordnance Survey maps

This is about what OS maps are, and what they show, and how to use them.

This OS map shows part of the Land's End peninsula in Cornwall.

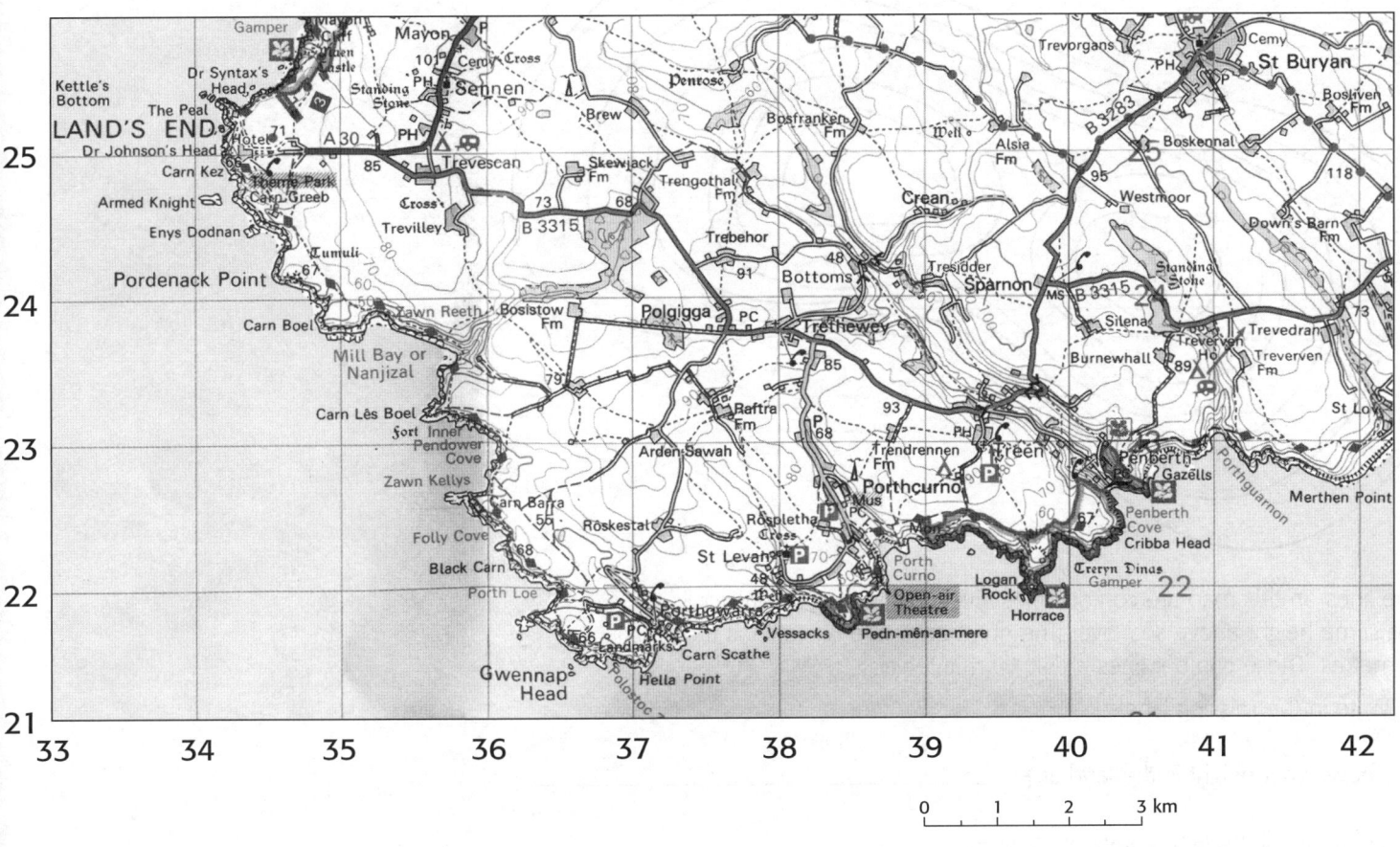

1 What is at this grid reference?

a 387219 _____

b 385253 (Hint: Fm means farm) _____

2 Find one of these on the map and give a six-figure grid reference for it.

a a car park _____

b a church _____

c a public phone _____

3 What clues are there on the map that the Land's End peninsula gets lots of visitors?
Give as many as you can.

_____

_____

2.9 ◄ **How high?**

This is about how height is shown on an OS map.

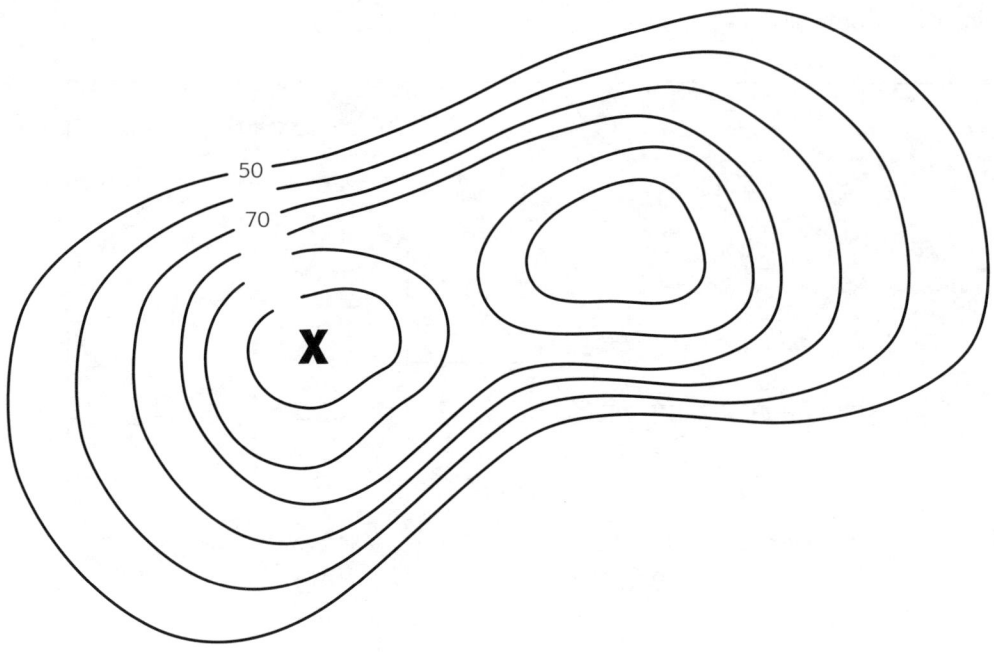

1 The lines on this map are contour lines. Everything along a contour line is the same height above sea level. The number on the line shows the height in metres. These contour lines are at 10 m intervals.

   a Write in the missing labels.

   b Above what height is the land at X? _____

   c Colour in all the land above 80 metres.

2 Now look at the OS map on page 15. About how high above sea level is:

   a Trevilley (358246)? _____

   b Raftra Farm (376233)? _____

3 a Complete this sentence:

    Another way that OS maps show how high a place is by using _____

    _____. These give the exact height at a spot, in

    metres above sea level.

   b Can you find an example of this on the OS map on page 15? Give a four-figure grid reference.

    _____

2 ► And to finish ...

Time yourself to check how well you know your map words! Tick the correct answer.

1 Walter has a connection with Hong Kong – his penfriend lives there. But what type of connection is this?

national ☐   local ☐   international ☐

2 You draw a simple map for a friend to show them how to get to your house. What's this kind of map called?

sketch map ☐   Ordnance Survey map ☐   easy map ☐

3 What are the lines on a map that show height above sea level called?

height lines ☐   contour lines ☐   confluence lines ☐

4 What term describes the straight line distance between two places?

as the crow flies ☐   as the kestrel hovers ☐   point-to-point ☐

5 What term describes a photo taken from the air?

air shot ☐   sky photo ☐   aerial photo ☐

6 You *won't* see one of these on a sketch map. Which one?

an N arrow ☐   a scale ☐   a key ☐

7 What term describes a set of numbers, or numbers and letters, that helps you find a place on a map?

grid reference ☐   scale ☐   Ordnance point ☐

3.1 ◀ Settling down

This is about the things people looked for, when choosing a place to settle in.

1 a The letters in these words are jumbled up. Write the correct words in the boxes.
Some letters have been written in already, just to get you going.

mettlesent   ties   attisuion

| _ i _ _ |

the position of a settlement in relation to features such as rivers, hills, and other settlements

| _ _ t _ _ _ _ _ _ |

a place where people live; it could be a hamlet, village, town or city

| s _ _ _ _ _ _ _ _ |

the land a settlement is built on

b Now draw a line linking the key words you have written to their correct meaning.

2 It is 500 AD. You and your tribe are looking for a good place to settle in. Which features below will you *avoid*? Cross them out, so that you're left with the 'good' ones.

Features of a good place to settle in

| exposed to wind and rain | steep land | trees for fuel | difficult to get out of | land that floods easily |

| flat land with good soil | nice view | just rock and sand | lots of materials nearby, for making things |

| shelter from wind and rain |

| no rivers or springs nearby | easy to spot enemies approaching | good clear water for drinking | easy access to other places for trading things |

3 Look at this photo. Write down three reasons why you think people chose this site to settle on.

- _____
- _____
- _____

3.2 ◄ **Example: settling in Aylesbury**

This is about who settled first in Aylesbury – and why.

1 Aylesbury is a town about 55 km from London. The first people to really settle here were the Saxons. This sketch map shows what they would have found when they arrived about 1500 years ago.

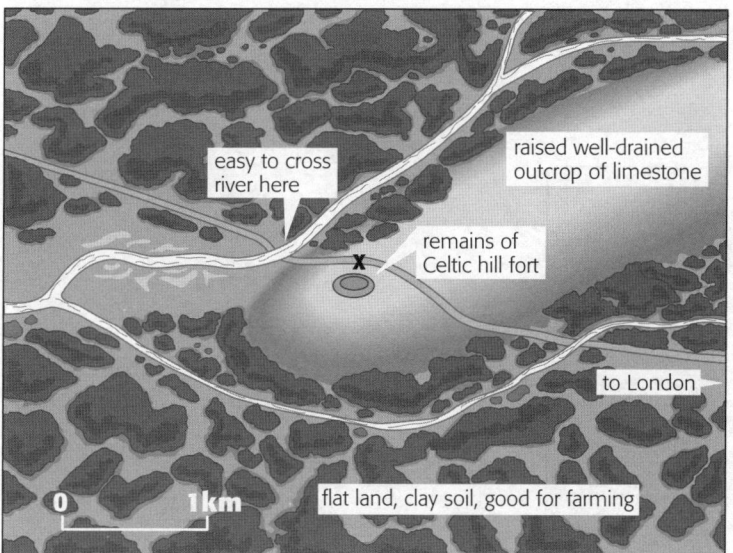

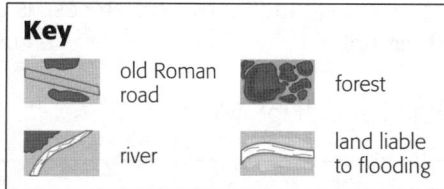

Key

old Roman road

forest

river

land liable to flooding

easy to cross river here

raised well-drained outcrop of limestone

remains of Celtic hill fort

to London

0   1km

flat land, clay soil, good for farming

They chose to settle at **X**. What reasons can you think of for this? (Think about access to water and to farmland for example.)

_____

_____

2 Settlements are all different shapes.

a Look at the boxes below. You need to match each settlement shape with the correct description and drawing. Colour each set of matching boxes a different colour.

nucleated

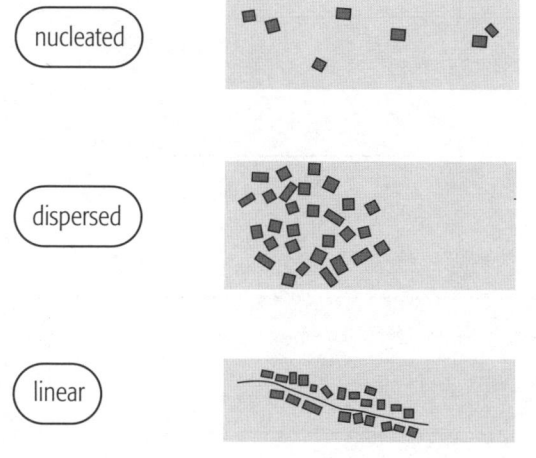

the buildings are strung out along a road

dispersed

the buildings are all spread out

linear

the buildings form a cluster

b Now have a look at the map on page 15. What shape is St Buryan (4025)?

_____

3.3 ◄ How Aylesbury grew

This is about how and why settlements grow.

1 Read this paragraph. Cross out the wrong words so that it makes sense.

If the position/site is a good/bad one, the settlement will grow. Small sites/settlements can grow into villages, then towns/hamlets, then towns/cities. As they grow, they take over the countryside or rural/urban areas around them. This process is called revolution/urbanisation. Situation/urban means 'built up'.

2 This graph shows population growth for Aylesbury.

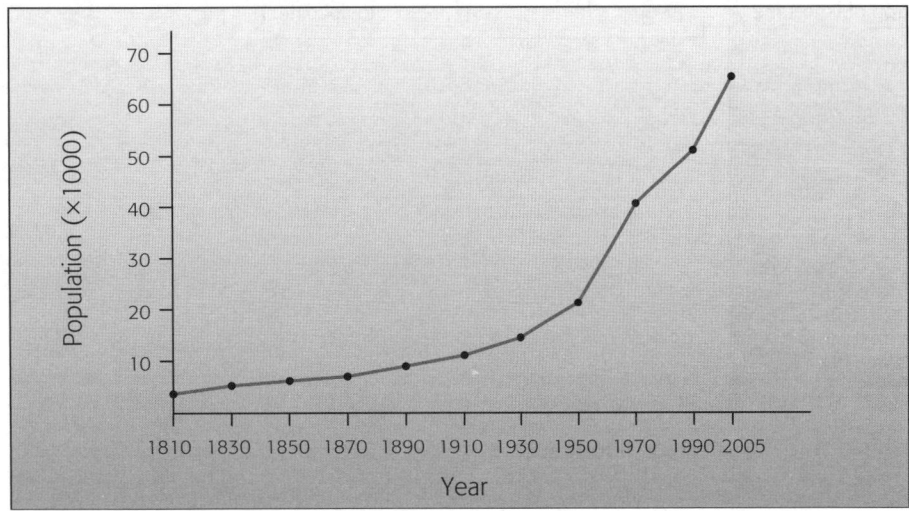

a Mark in these three events on the graph.
  1 1839 Birmingham Railway links it to Birmingham
  2 1865 Hazell, Watson and Viney, printers from London, set up a factory
  3 1991 New shopping centre opens

b How do you think each event helped Aylesbury to grow?

Event 1:

_____

Event 2:

_____

Event 3:

_____

3 Give four factors that would help a settlement grow.

_____   _____

_____   _____

3.4 ◄ The pattern of growth

This is about the pattern of factories, houses, and offices in our towns and cities.

1   Are these statements **True** or **False**? Answer by ticking the correct box.

                                                                                  True   **False**

    a A settlement usually grows out from the centre – so that's where the newest buildings are.   ☐   ☐

    b As the settlement grew, homes in the centre were turned into shops and offices, which people could reach easily from all directions.   ☐   ☐

    c So the main shops and offices are still in the centre. It's called the **central shopping district** or **CSD**.   ☐   ☐

    d The first factories were built along canals, rivers or railways, so that goods could be moved easily.   ☐   ☐

    e Rows of small cheap terraced houses were built close to the factories, for the workers.   ☐   ☐

    f As the population grew, new houses were built at the edges of the settlement, where land was cheaper.   ☐   ☐

    g Today, new industries are usually set up in the centre of town.   ☐   ☐

2   For each statement you said was false, write down how it should be written.

    _____

    _____

    _____

3   **a** Fill in the gaps in the boxes below to label this urban model.

    b Draw a line from each box to the correct part of the diagram.

    c Now colour each box and its part of the diagram the same colour. You will need six different colours.

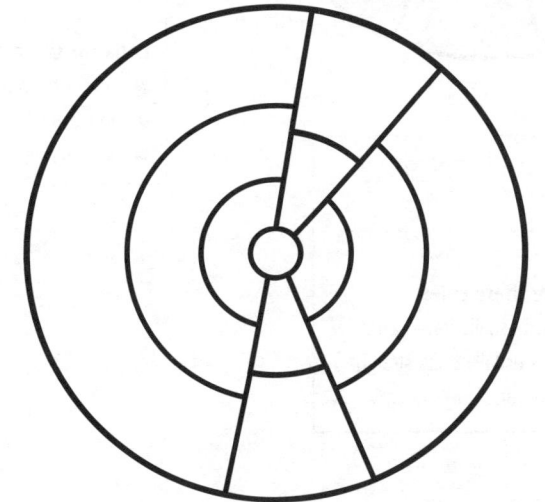

New industrial area
Industrial estates and business parks built since 1970, close to main roads.

Old i_____ area
Along a river, canal or railway. Many old factories now closed. Area may look run down.

The _____
Large shops and offices. Restaurants, cafes, museums, cinemas and theatres.

Modern housing
New houses and housing estates. New shopping centres. Parks and other open areas. This area is the **o_____ s_____**.

19th century housing
Mostly terraced houses for factory workers. Some now replaced by high-rise flats. Small corner shops nearby. This area is the **inner city** or **t_____ z_____**

Housing 1920–1950
Larger houses usually with gardens. Some parks. Some rows of shops. This area is the **i_____ s_____**.

3.5 ◄ Be a land-use detective!

This is about how to identify different types of land use on an OS map.

a First, use the clues in the boxes below to work out the type of land use. One box has been filled in for you. Choose from:

- The central business district
- An old industrial area
- New industrial area
- An area of 19th century workers' housing
- An area of modern housing

b Then, draw lines to show where each area fits into the urban model.

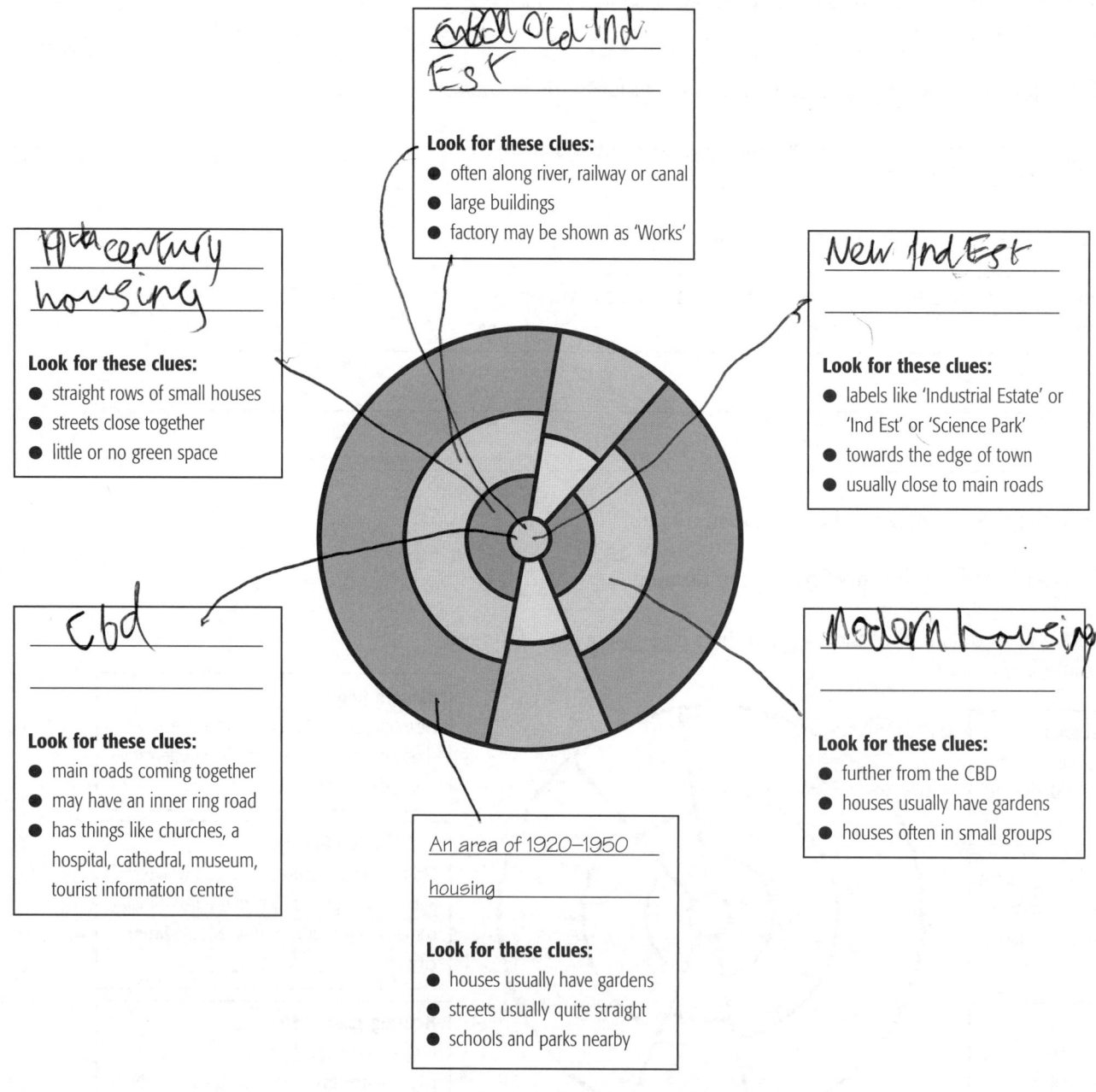

Old Ind Est

Look for these clues:
- often along river, railway or canal
- large buildings
- factory may be shown as 'Works'

19th century housing

Look for these clues:
- straight rows of small houses
- streets close together
- little or no green space

New Ind Est

Look for these clues:
- labels like 'Industrial Estate' or 'Ind Est' or 'Science Park'
- towards the edge of town
- usually close to main roads

CBD

Look for these clues:
- main roads coming together
- may have an inner ring road
- has things like churches, a hospital, cathedral, museum, tourist information centre

Modern housing

Look for these clues:
- further from the CBD
- houses usually have gardens
- houses often in small groups

An area of 1920–1950 housing

Look for these clues:
- houses usually have gardens
- streets usually quite straight
- schools and parks nearby

3.6 How's Aylesbury doing today?

This is about the good and bad things people think about where they live.

1 a Look at the boxes below which show some features that may be found in a settlement. Colour in the boxes that you think show good things about a place.

department store   park   prison   airport   leisure centre

college   railway station   shopping centre   main road

nightclub   river   cycle lanes   industrial estate   housing estate

b Pick one good thing and one thing that you would grumble about. Explain your feelings.

Good feature:

_____

Reason:

_____

_____

Grumble:

_____

Reason:

_____

_____

2 In the box below, design and draw a logo or advert for your own settlement.

3.7 ◀ A new challenge for Aylesbury

This is about why the UK needs more homes – and the conflicts over where to put them.

1 Tick the correct answer.

a About how many new homes does the UK need each year?

2400 ☐   24 000 ☐   240 000 ☐

b By 2020 the population of Aylesbury is expected to rise by…

3000 ☐   30 000 ☐   300 000 ☐

c How many homes does the UK have to build by 2020?

3 thousand ☐   3 million ☐   30 million ☐

2 Cross out the incorrect word in these sentences.

a More/fewer people are getting divorced.

b Some houses are so big/run-down that people don't want to live in them.

c More/fewer people are living on their own.

d It's easier/harder to build on greenfield sites.

e It's cheaper/more expensive to build on brownfield sites.

3 Look at this advert. Then answer the questions below.

a What type of development is the Millside?

brownfield ☐   greenfield ☐

> **The Millside**
> Luxury two- and three-bedroomed apartments.
> Fully-equipped kitchens. Parking space.
> Five minutes' walk to city centre.

b What sort of people do you think would like to live in a development like this?

_____

4 Find out about one housing development near you. Then make brief notes about it under these headings.

Name and location of housing development

_____

Number and type of houses being built

_____

What the land was used for before

_____

Who's for it, and who's against it

_____

3.8 ‹ Sustainable development for Aylesbury

This is about how sustainable towns and cities are.

1 Look carefully at the photo below and the four 'points of the compass' around it. In the spaces around the photo write questions that you would like to ask about the settlement.

Natural
This is about the environment – energy, air, water, soil, living things.

_____

_____

Who decides
This is about who makes choices and decides what is to happen.

_____

_____

Economic
This is about money, trade, buying and selling.

_____

_____

Social
This is about people and the way they live their lives.

_____

_____

To help you remember these four points, think; **N**orth, **E**ast, **S**outh, **W**est.

2 Do you think that the settlement shown is sustainable? Yes/No (Circle your answer)
Give one reason why.

_____

_____

3 Write down two ways in which you think the place could become more sustainable.

_____

_____

_____

3 And to finish ...

Hidden in this wordsearch are 12 key words to do with settlement.

Find the words in all directions (you can circle or colour them).
If you need clues, look at the sentences below.

f	e	s	k	u	f	l	e	l	b	a	n	i	a	t	s	u	s
y	d	d	e	t	a	e	l	c	u	n	l	d	n	t	k	a	s
e	a	f	s	u	d	h	g	r	m	f	l	o	s	f	f	t	k
u	r	b	a	n	r	e	g	e	n	e	r	a	t	i	o	n	s
t	s	r	s	o	s	f	r	j	m	v	p	s	e	g	j	e	d
t	r	o	x	i	w	e	e	u	y	d	i	k	d	r	e	m	s
e	j	w	c	t	y	h	e	c	r	m	u	s	e	b	c	e	b
r	r	n	v	a	e	g	n	g	d	a	e	d	v	d	s	l	c
r	v	f	m	u	a	a	f	v	c	n	l	f	f	a	s	t	h
a	v	i	j	t	d	r	i	v	k	g	w	a	u	b	n	t	t
c	j	e	g	i	c	r	e	e	r	e	q	m	r	s	x	e	d
e	s	l	d	s	j	q	l	q	e	s	q	b	b	e	b	s	e
d	t	d	m	r	k	f	d	y	c	k	s	v	a	j	a	e	j
l	a	g	a	d	g	f	k	h	l	v	g	f	n	n	k	r	f
z	q	b	m	s	h	v	c	b	d	n	v	v	a	e	l	f	m
t	t	n	e	m	p	o	l	e	v	e	d	e	r	u	t	d	l
k	l	s	f	j	i	o	y	u	w	f	m	z	e	j	l	i	f
u	d	i	s	p	e	r	s	e	d	a	r	h	a	m	k	o	s

A A _____ site is a site which has not been built on before.

B A _____ is where people live, for example a town.

C A _____ _____ is the countryside.

D In a _____ settlement, the buildings are all spread out.

E A _____ is the land a settlement is built on.

F _____ is the position of a settlement in relation to other features such as rivers, hills and other settlements.

G The _____ is the short name for the centre of a town or city.

H In a _____ settlement, the buildings form a cluster.

I An _____ _____ is a built-up area.

J _____ is when an area is rebuilt for a new use.

K _____ _____ is when a run-down urban area is redeveloped and brought to life again.

L A _____ way of life means not wasting things or harming the environment.

M A _____ site is one that was built on before.

4.1   **Shopping around**

This is about why shops are where they are – and how shopping is all tied up with geography!

1 a Fill in the gaps in these sentences.

Things you shop around for, and compare in different shops, are called

_____ goods.

Low-cost goods which you buy in the nearest convenient place, are called

_____ goods.

b For each type of goods, give two examples that you or your family might buy.

_____

_____

2 a Look at the boxes below. Four contain settlements with their populations and the others contain a range of shops. Which shops would you expect to find in settlement A? Colour in those boxes and the box for settlement A in the same colour.

b Match the other boxes. Colour each set a different colour.

Settlement A	**Settlement B**	**Settlement C**	**Settlement D**
Population: 70 000	*Population*: 125	*Population*: 1800	*Population*: 8000

Large new shopping centre

1 newsagent

No services

1 small supermarket

2 clothing boutiques

9 shoe shops

All the main chain stores including Monsoon and Next

1 large supermarket

A video rental shop

c Now fill in these gaps using words from the box below.

The _____ the settlement, the _____

choice it offers for shoppers, since shops will be attracted by the larger

_____ base.

larger	smaller	choice
customer	less	more

4.2 Out-of-town shopping: Bluewater

This is about Europe's biggest out-of-town shopping centre, and its impact.

You can explore:
● over 330 stores
● 40 cafés, bars and restaurants (including internet cafés)
● a 13-screen cinema
● a climbing wall
● six boating lakes
● bike hire – and 50 acres of parkland to cycle round
● a golf putting course
● a fishing lake
● a discovery walking trail

1 Bluewater is the largest out-of-town shopping centre in Europe. Have a look at the list above to see what it has to offer.

a What are the advantages of such mega-centres for the customer? (Try to think of five) Think about the whole family in your answer.

_____

_____

_____

_____

_____

b Give three reasons why developers choose to build mega-centres on out-of-town sites.

_____

_____

_____

2 The UK has 11 mega-centres, and the government has decided not to allow any more. What harmful effects might mega-centres have on the environment?

_____

_____

4.3 ▸ Shopping on the internet

This is about how internet shopping works – and the pros and cons for different groups of people.

1 Internet shopping is the latest big change in shopping. To do it, you have to have a computer that can log on to the internet. You can then shop online, and the goods will be delivered to your door.

a List four advantages of internet shopping for the customer. Hint: think about the sort of people who can't get to the shops easily, as well as those who can.

_____

_____

_____

_____

_____

_____

b Books were one of the first things sold on the internet. Why has online bookselling been so successful?

_____

_____

_____

2 On the shopping list below, circle in *green* the things people might buy on the internet, and in *red* the things that they probably wouldn't.

Shopping list
a Kylie CD   a computer game
cream cakes   a pair of jeans
a geography workbook   chocolate bar
a second home by the sea

3 In the space provided here, design a 'pop-up' (a box that appears on the internet with an advertisement in it) for your online book company. You need to give the advantages of buying books from your company … and say that your books are cheaper than at the bookshop in town!

4 ◀ And to finish …

What have you learnt about shopping? Let's find out. Tick the correct answer.

1 One of these is a shopping mega-centre. Which one?

Greenpond ☐   Bluewater ☐   Lakeside ☐

2 Complete the sentence. Goods like clothes and shoes, where you like to see a choice before you buy, are …

convenience goods ☐   selection goods ☐   comparison goods ☐

3 Bluewater is built on the site of an old chalk quarry. What term best describes this kind of site?

brownfield ☐   bluefield ☐   greenfield ☐

4 Which of these would you be *least* likely to buy on the internet?

a pet kitten ☐   a CD ☐   a holiday ☐

5 Which term is this an answer to? *What you charge your customers for the goods, minus what you paid for them.*

profit ☐   costs ☐   expenses ☐

6 Which of the following goods would you describe as convenience goods?

a new atlas for your geography ☐   the daily newspaper ☐   a new pair of trainers ☐

7 Companies that buy land and put up buildings for rent or sale are called …

developers ☐   estate agents ☐   architects ☐

8 Which is a village *least* likely to have?

a newsagent ☐   a bakery ☐   a department store ☐

5.1 ◄ Your island home

This is about the forces that shaped the British Isles – and about Britain's main physical features.

1 This paragraph explains how our island home became an island. Choose words from the box to fill in the gaps.

Once upon a time, the British Isles lay at the _____, as part of a giant

_____. When this broke up, they drifted _____ as part of Europe.

As they drifted, over millions of years, they went through many _____.

They became desert. They were frozen in _____. They were drowned by

the _____. They had earthquakes and eruptions. They got pushed and

squeezed until _____ grew. And then they got _____ from the

rest of Europe.

ice	cut off	changes	sea
crust	continent	mountains	currents
north	equator		

2 Here are some features of the British Isles, but they're all jumbled up. Unscramble the words and then use them to label the map below.

vrier sernve   verri nttre
rierv htmaes   alke drictist
eisglhn nanchel   ninespen
rthno aes   rthno estw landshigh
isirh sae

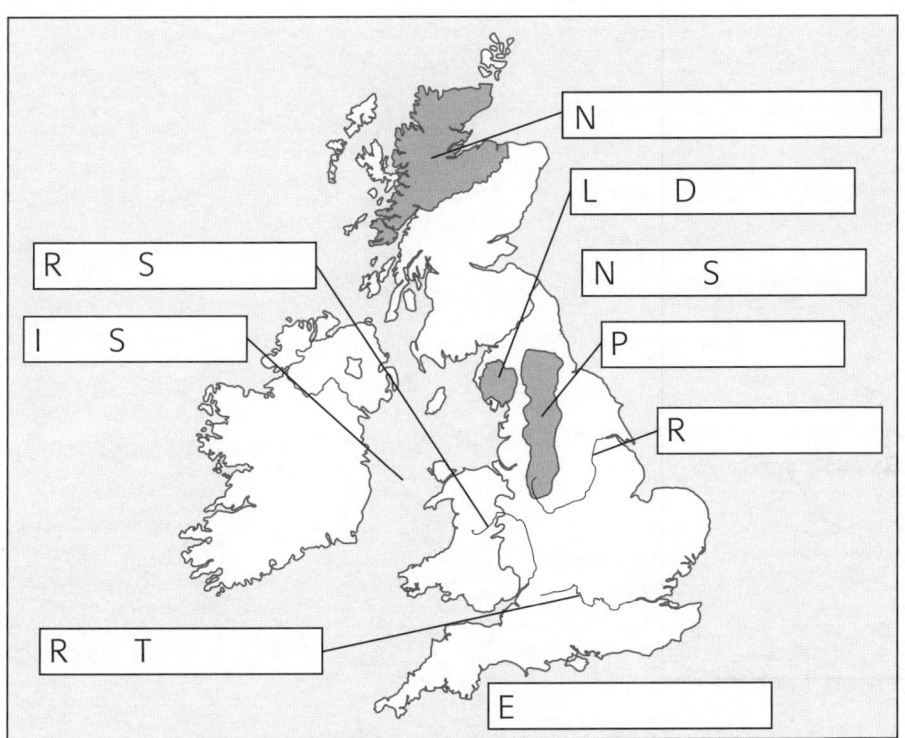

5.2 It's a jigsaw!

This is about how we humans have carved up the British Isles.

1 Fill in the gaps using words from the box.

The British Isles is divided up into two countries: the United Kingdom and the

_____ .

The _____ in turn is made up of different nations: England,

_____ , Wales and Northern Ireland.

United Kingdom	Germany	British Isles
Republic of Ireland	Scotland	England

2 Now look at the map and answer these questions.

   a A is called _____

   b B is called _____

   c D is called _____

   d A–D together are called _____

   e A–E together are called _____

   f Finally, shade in the countries that make up Great Britain.

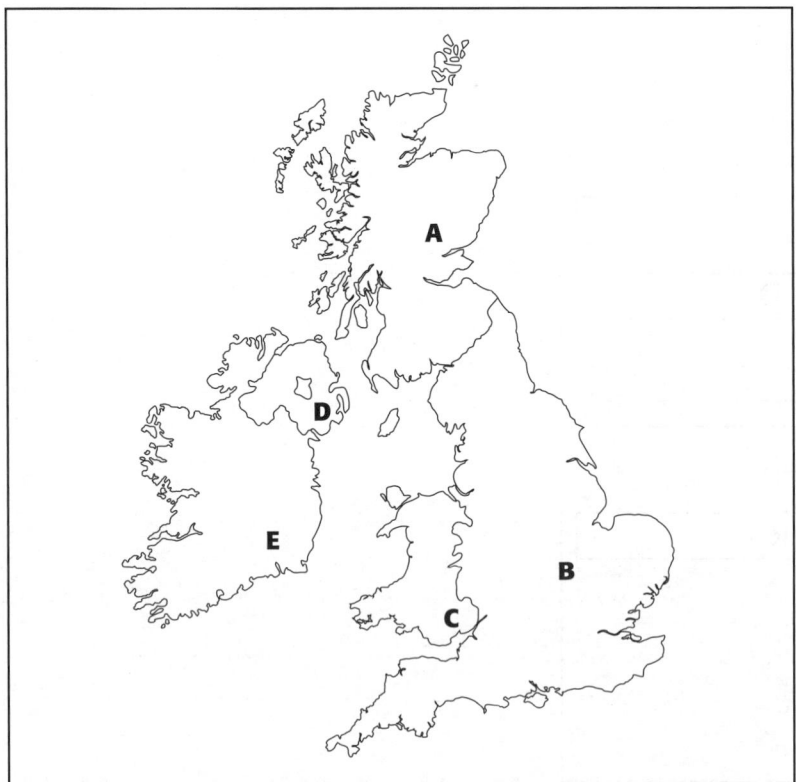

5.3 ◄ **What's our weather like?**

This is about the difference between weather and climate – and how the climate varies across the UK.

1 Fill in the gaps.

● _____ means the state of the _____. Is it warm? wet? windy?
  It changes from day to day.

● _____ is the _____ weather in a place.

weather	average	atmosphere	climate

2 a Cross out the wrong word in these sentences.
   In general:

● It is colder/warmer in the north, because it is further from the equator.
● It is also colder/warmer on high land. Up a mountain the temperature falls/rises.
● But in winter, a cold/warm ocean current called the North Atlantic Drift cools/warms
  the west coast. So the east coast is the coldest/warmest part in winter.

b The maps below show average temperatures in summer and winter. Colour in the
   first map in shades of orange. Make the warmest areas darkest, and the coldest
   areas lightest.

c Now colour the second map in blue. This time, make the coldest areas darkest.
   And don't forget the key!

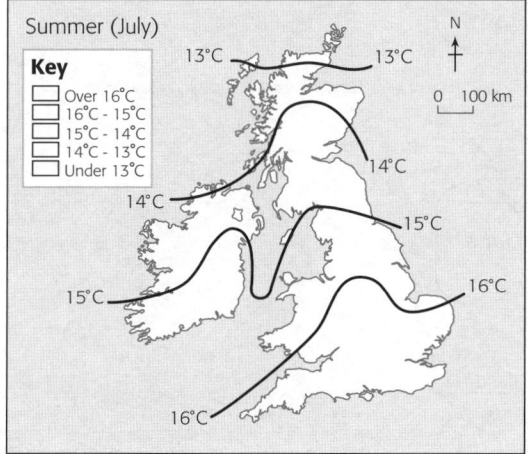

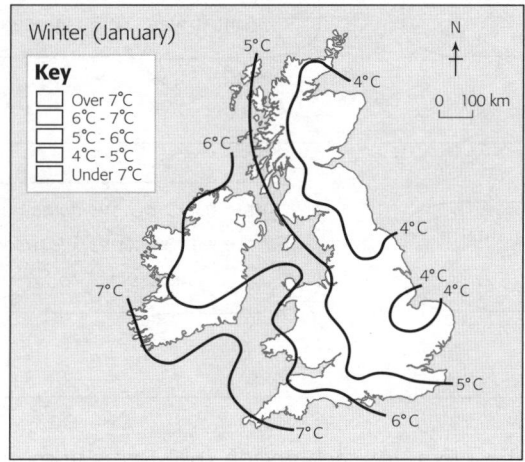

3 Answer these questions in full sentences.

a Which parts of the British Isles are wettest?

_____

b Can you explain why?

_____

_____

5.4 ◀ Who are we?

This is about how Britain has been peopled by immigrants.

1 Some of the definitions below are incorrect, but which? Cross out any wrong terms and write the correct word in the second column. One has been done for you.

	Correct term
A An **asylum seeker** is a person who flees to another country for safety, and asks to be allowed to stay there.	
B An **invader** is someone who enters a country to attack it.	
C An ~~emigrant~~ is a person who comes into a country to live.	immigrant
D A **settler** is a person who takes over land to live on, where no one has lived before.	
E A **refugee** is a person who has been forced to flee from danger.	
F An **asylum seeker** is a person who leaves his or her own country to settle in another country.	
G A **refugee** is a person who moves to another part of the country or another country, often just to work for a while.	

2 Here are some statements from people who've arrived in the British Isles in the last 2000 years.

> It's 48 AD. I am a centurion with the Roman army. Our aim is to expand our empire.

> I came to live here a few years ago, in 2001. I didn't want to leave Kosovo but the war meant it was too dangerous to stay.

_____

_____

> It's 1956. I've come here from Jamaica in search of a job.

> It's 4000 BC. I've come here from Europe with my tribe. We're looking for a good place to farm.

_____

_____

Choose what you think is the best term for each person (use terms from question 1).

3 Do you agree with this statement: 'In the British Isles, we are all immigrants'?

_____

_____

_____

5.5 Where do we live?

This is about how we humans have shaped the country, through where we chose to live!

1 Fill in the gaps.

The _____ _____ of a place is the average number of people per square kilometre.

2 Tick the correct answer.

 a The nation with the highest population density is …

England [ ]   Wales [ ]   Scotland [ ]

 b Of these areas, the one with the lowest population density is …

Cumbria [ ]   Greater Manchester [ ]   Devon [ ]

 c Of these cities, the largest is …

Edinburgh [ ]   Birmingham [ ]   Glasgow [ ]

3 Look at this pie chart for the United Kingdom

Where the UK population lives

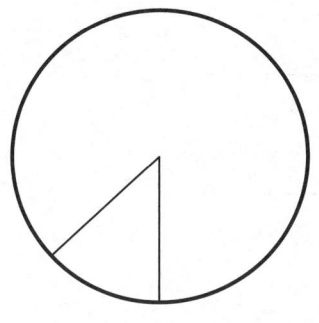

Key
[ ] urban areas
[ ] rural areas

 a Shade in the chart and the key to show where the population of the UK lives.

 b Imagine you live on a farm half an hour's drive to the nearest town.
Give three good points and three bad points about living in such a rural area.

Good points

_____

_____

_____

Bad points

_____

_____

_____

5.6 ◀ What kind of work do we do?

This is about the kind of work people in the UK do for a living.

1 What is *economic activity*?

_____

2 a Cross out the wrong word in each sentence.

The secondary/primary sector is when you gather materials from the earth.

The secondary/quaternary sector is when you do hi-tech research.

The primary/tertiary sector is when you provide a service for people.

The service/manufacturing sector is when you turn materials into things to sell.

b Now give two example of jobs in each sector.

Primary:

_____

Secondary or manufacturing:

_____

Tertiary or service:

_____

Quaternary:

_____

3 In the UK, about 26 million people work for a living. This pie chart shows the kind of work we do in the UK.

a Shade in the boxes in the key, each a different colour.

b Now shade in the different sections of the pie chart to show the kind of work we do.

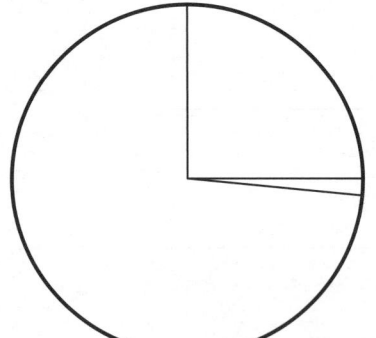

Key
- ☐ manufacturing
- ☐ services
- ☐ primary

5.7 ◄ **High or low earnings?**

This is about how some parts of Great Britain are better off than others – and some of the reasons why.

1 What helps to make an area wealthy? Cross out the things that don't help, so that you're left with the things that do.

Things that help make a place wealthy

(good farmland)   (oil (or other valuable mineral))   (companies offering poorly-paid jobs)

(few transport links)   (beautiful countryside)   (interesting history)

(land that is not good for farming)   (little industry)   (museums and other tourist attractions)

(flourishing industry)   (no tourist attractions)   (plenty of companies offering highly-skilled and highly-paid employment)

(easy access (having good road links for example))

2 This is Anya. She works at a large museum which attracts lots of tourists. It was set up a few years ago by a grant from the government.

Fill in the speech bubble to say how the museum has affected Anya and the local area.

Three years ago, I had no job and very little money. It was really tough. I thought about moving away because the area felt so run down. It was

_____

_____

Now things are different. What happened was

_____

_____

_____

I feel _____

_____

The local area has changed too …

_____

_____

_____

_____

5.8 ◄ The UK in the world

This is about how you and the UK are linked to other places around the world.

1 This paragraph describes some of the links the UK has with other countries. Choose words from the box to fill in the gaps.

English is the main _____ used for business around the world. The UK

_____ all over the world and outside the EU its main trading partner

is the _____. The UK is linked to other countries through _____,

and over 30 million trips are made to the UK from people from other countries. Many

British companies have branches all over the world. They are called _____

companies. To look after the environment, the UK has signed many _____

_____ with other countries and in 2006 it spent almost £7 billion in aid

to help _____ countries.

transnational	USA	tourism	services
language treaties	poorer	groups	trades

2 Why do you think the UK gives aid to other countries?

_____

_____

3 Tourists spend a lot of money in the UK each year. Explain how this helps people who live and work in the UK?

_____

_____

_____

_____

5 ◄ And to finish ...

Time to test your understanding of some key words from the *Exploring Britain* chapter!

Use these clues to complete the crossword.

Across

2 This river flows by the Houses of Parliament
4 Made up of England, Scotland, Wales and Northern Ireland
10 Something set up to meet people's needs, for example a shop or a school
13 To do with the countryside
14 The average weather in a place
15 One of the four nations that makes up the United Kingdom
16 15 countries in the EU share this currency

Down

1 A person who has been forced to flee from danger
3 A built-up area, such as a town or city
5 A person who comes into a country to live
6 When you turn materials into things to sell
7 The dry area on the leeward side of a hill
8 The part of the economy where people take things from the earth and sea
9 One of the four nations that makes up the United Kingdom
11 The United Kingdom is made up of four of these
12 In this part of the economy, people provide services

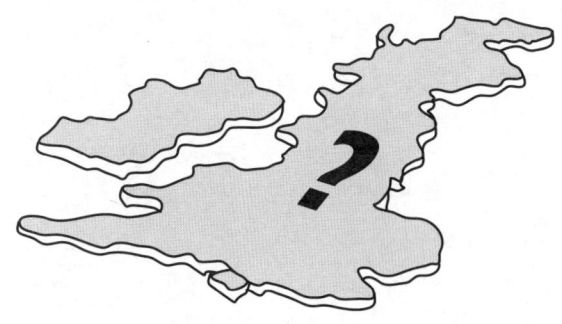

6.1 ◀ **The water cycle**

This is about the water cycle, and how rainfall reaches a river.

1 Water moves between the ocean, the air and the land. This circulation is called the water cycle.

    a Fill in the gaps below, choosing words from the box (you don't have to use them all).

       ☐ The air _____. High up, where it's cooler, the water vapour _____ into tiny water droplets. These form _____.

       ☐ The water drops fall as rain (or hail or sleet or snow).

       ☐ The sun warms oceans, lakes and seas, turning water into water vapour. This is called _____.

       ☐ Some water runs along the ground, and some soaks through it, heading for streams and rivers.

       ☐ The droplets inside the clouds grow into larger droplets, leading to _____.

       ☐ The river carries the water back to the _____. The _____ is complete. And then it starts all over again…

evaporation	rises	infiltration
gas	precipitation	clouds
condenses	ocean	cycle

    b Add numbers in the small boxes so that the sentences are in the correct order.

2 Draw a diagram in the box below to show how rainwater reaches a river. Try to use as many of these words as possible (there are a few clues to help you!):

interception (when rainwater catches leaves)   groundwater flow
surface runoff   infiltration (the soaking of rainwater into the ground)
throughflow   permeable (lets water soak through)
groundwater   impermeable

6.2 ◀ **A river on its journey**

This is about the different parts of a river.

1 a Cross out the incorrect word in these sentences.

- The point where two rivers join is called a tributary/confluence.
- The confluence/watershed is an imaginary line that separates one drainage basin from the next.
- The source/mouth is where the river flows into a lake, or the sea, or the ocean.
- The flat land around a river that gets flooded when the river overflows is the flood plain/tributary.
- The mouth/source is the starting point of the river.
- The land around a river from which water drains into the river is the drainage basin/watershed.

b Now draw a sketch map of an imaginary river. Try to mark on and label all the features from **1a**.

2 Fill in the gaps.

A drawing of the river's _____ _____

The slope gets less steep in this middle stretch.

The _____ is the highest point.

The _____ is the river's lowest point.

lake or sea

different layers of rock below the river

Now the slope is flattening out.

6.3 ◄ **Rivers at work**

This is about how rivers shape the land, by picking up, carrying and dropping material.

1 Fill in the gaps choosing words from the box.

Rivers do their work in three stages:

1 They pick up or _____ material from one place.

2 They carry or _____ it to another place

3 Then they drop or _____ it.

| erode | transport | deposit |

2 Finish off this cartoon to tell the story of Sid the Stone's journey. (You don't have to fill all the boxes.)

1

Sid the stone had lived in the river bank for as long as he could remember. Then, one day …

2

… he was prised out of the bank by **hydraulic action**!

3

4

5

6

6.4 ◀ Landforms created by the river

This is about the landforms a river creates, by eroding and depositing material.

1 Fill in the gaps in this table.

Landform	Description	Created by ...
V-shaped valley	a valley shaped like the letter V, carved out by a river	
waterfall		erosion
gorge	a narrow valley with steep sides	erosion
	a bend in a river	erosion + deposition
oxbow lake	a lake formed when a loop of river gets cut away	

2 a These pictures show how a waterfall develops.
Under each picture, describe what is going on.

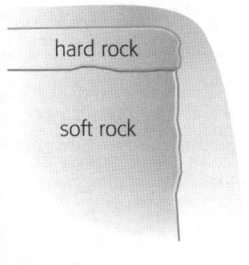
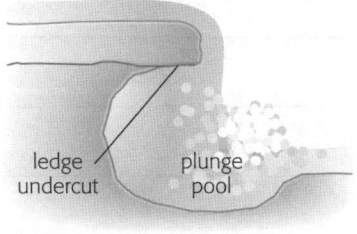
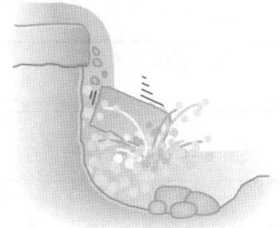
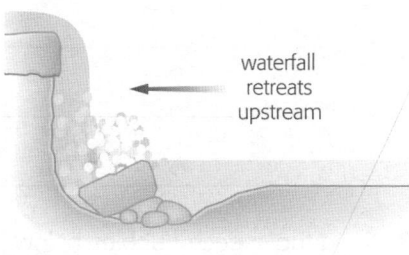

1 _____

2 _____

3 _____

4 _____

b Draw pictures in the boxes below to show how a meander develops.

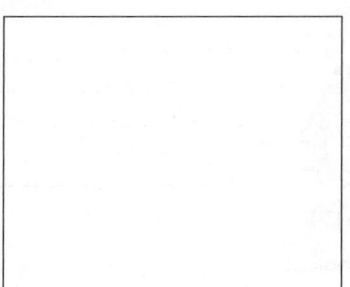
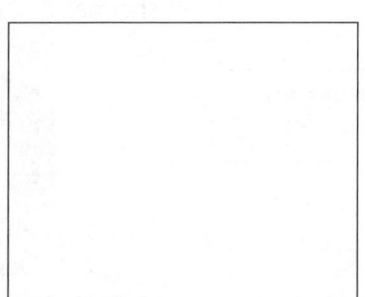
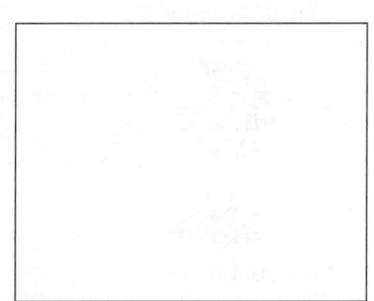

1 Water flows faster on the outer curve of the bend, and slower on the inner curve. So …

2 … the outer bank gets eroded, but material is deposited at the inner bank. Over time …

3 … as the outer bank wears away, and the inner one grows, a meander forms.

4 As the process continues, the meander grows more 'loopy'.

6.5 ◄ **Rivers and us**

This is about how we make use of rivers and how sometimes we may damage them.

1 a Look at the statements below. Draw a line linking the statements that you think are connected. One has been done for you.

A Dams are built across some rivers	And then pumped to our taps
B Farmers pump water from rivers	And then sent back to rivers
C Water from rivers is cleaned	To irrigate their land
D Dirty water from our houses is cleaned	So that the water turns turbines to make electricity

b Choose one of your completed statements from A to D. Copy it out on the line below.

_____

c Do you think what you have just written is good or bad for the river? Give reasons for your answer.

_____

_____

_____

2 Read the speech bubbles below.

Environmentalist

Rivers are for wildlife more than for people

Fisherman

My fishing lines get caught by the boats and break

Farmer

Without the water from the river my crops will not have enough water

Boat owner

We are quiet and do not disturb anybody

Dog walker

There are very few places left where we can walk our dogs in peace

Choose the one person that you think may damage the river the most. Give reasons to explain your choice.

_____

6 ◄ And to finish ...

All of the answers to the puzzle are key words to do with rivers.

1 Use the clues to complete the puzzle.

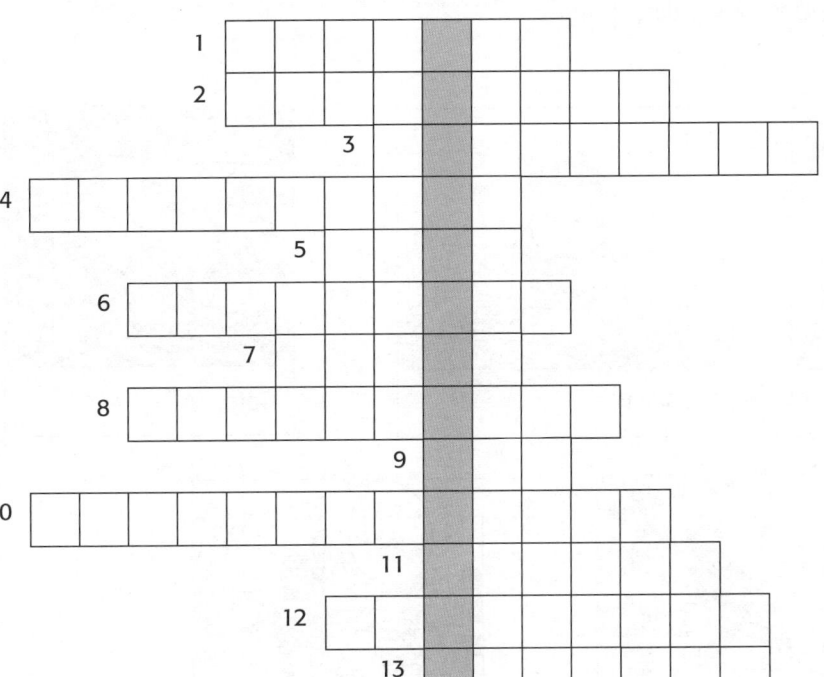

Clues

1 a bend in a river

2 an imaginary line separating one drainage basin from the next

3 where a river or stream flows over a steep drop

4 flat land around a river that gets flooded when the river overflows

5 the side of the river channel

6 formed when a loop in a river gets cut off

7 a narrow valley with very steep sides

8 where two rivers join

9 the base of the river channel

10 a feature shaped like the letter V, carved out by a river

11 the starting point of a river

12 a river that flows into a larger one

13 a waste product that gets into rivers and helps algae to grow

2 a What is the term in the shaded boxes?

_____

b What does this term mean? _____

_____

7.1 Tewkesbury under water

This is about the 2007 floods that hit Tewkesbury in south west England.

1 Look at the cartoons below and the list of dates. They show what happened to Ellen in the floods but are jumbled up. Sort them out and write the correct date in the space below the cartoon.

Date:

Date:

Date:

Date:

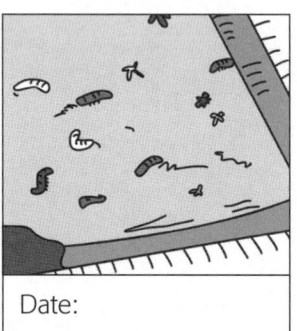

Date:

Date:

Date:

| 20th July | 21st July | 22nd July | 23rd July |

| 25th July | 27th July | 30th July |

2 Choose the day when you think Ellen would have been the saddest. Explain why you chose that day.

Date:_____

Reason: _____

_____

3 Write a short letter to Ellen giving her some advice on how to cope with the flooding.

_____

_____

_____

7.2 ‹ **What causes floods?**

This is about what floods are, what causes them – and how we humans can make them worse.

1 What are floods? Explain in one sentence.

_____

_____

2 There are lots of reasons why floods happen. These pictures show some of them.

a Draw round all the *natural* factors in one colour and the *human* factors in another.

b Now choose one reason. Fill in this writing frame to explain how it increases the risk of flooding. Add any suggestions you can think of that would reduce this risk.

_____ can increase the risk of flooding.

This is because _____

_____

_____

To make floods less likely this could be changed by _____

_____

_____

7.3 ▶ So – why did Tewkesbury flood?

This is about the reasons for the flooding that damaged Tewkesbury.

1 A sketch map of Tewkesbury has been started for you. Your task is to finish the sketch map and add some labels showing why the town flooded so badly in 2007. One has been done for you.

Don't forget to finish the key!

Reason 5

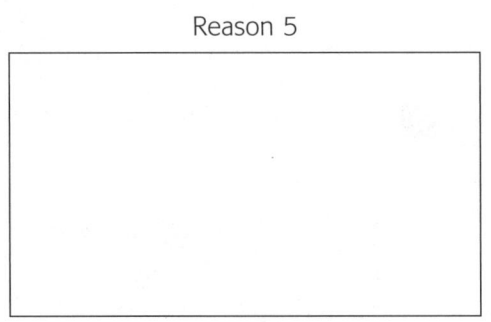

Reason 1

Tewkesbury is located where two rivers meet

Reason 2

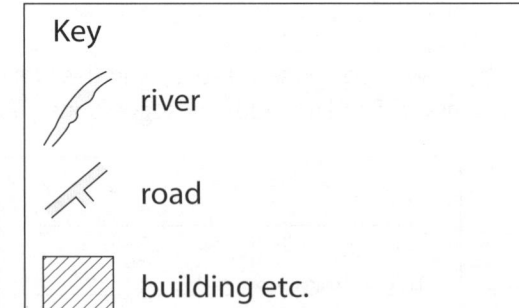

Key

river

road

building etc.

Reason 4

Reason 3

7.4 ‹ Who helps in a flooding crisis?

This is about the different groups who may help when flood disasters happen.

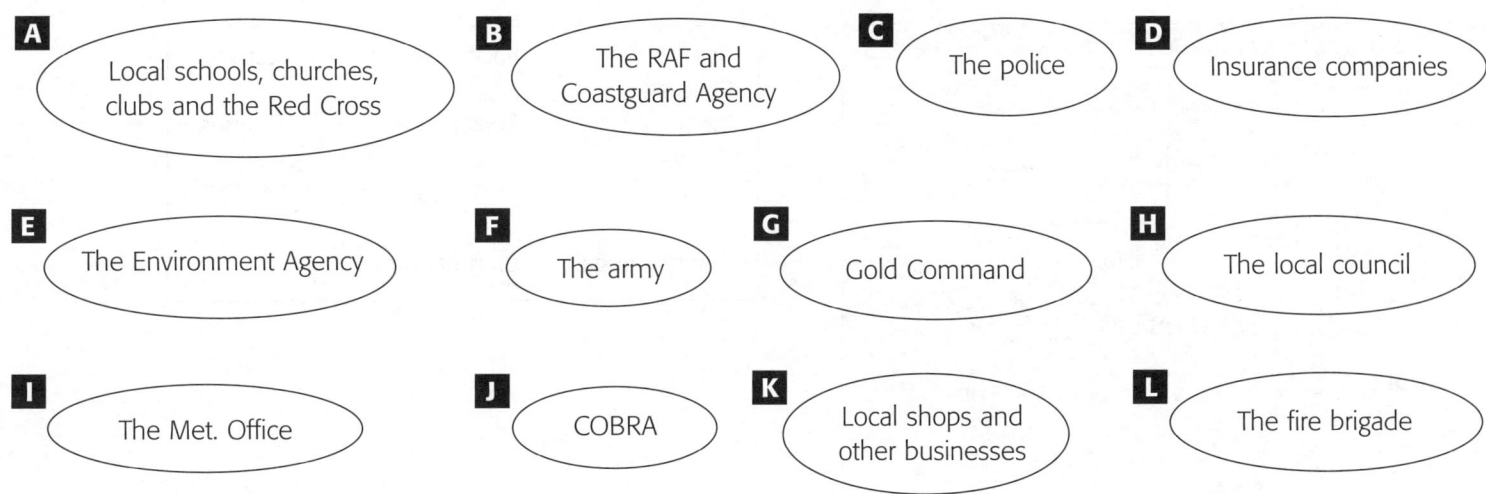

A Local schools, churches, clubs and the Red Cross

B The RAF and Coastguard Agency

C The police

D Insurance companies

E The Environment Agency

F The army

G Gold Command

H The local council

I The Met. Office

J COBRA

K Local shops and other businesses

L The fire brigade

1 These groups of people all offer help when floods happen, but often at different times.

On the line below write each letter in the place showing *when* you think they would be asked to help – before the flood happens, immediately the flood happens, later on or somewhere in between. One has been done for you.

/_____/_____**F**_____/_____/
    Before the flood   During the flood   After the flood

2 Pick one of the groups listed. Imagine that you are working with them when a flood happens. Describe what you did.

Group _____

This is what I did _____

_____

_____

_____

_____

_____

_____

7.5 ◢ **Flooding: the consequences**

This is about how floods can affect us all – even if we are nowhere near them.

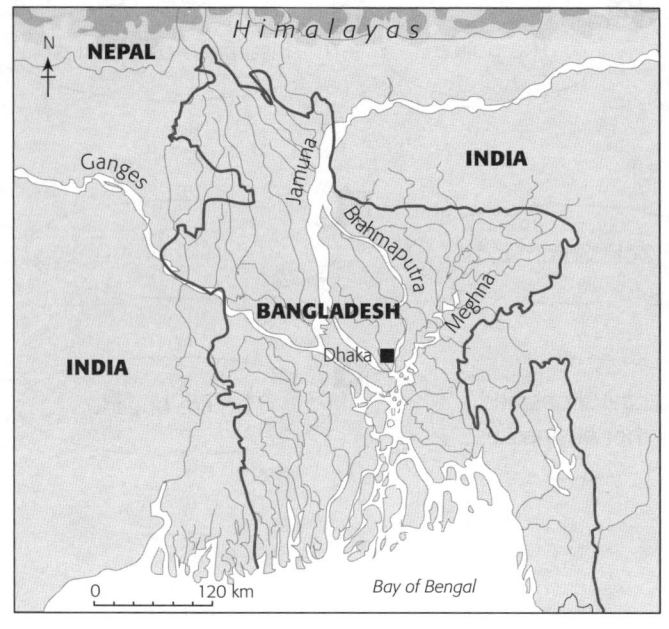

Asia	Pacific Ocean	Nepal
Pakistan		
India	Kolkata	Thames
Ganges		
Rhine	Himalayas	Dhaka
Alps		
Bay of Bengal	Brahmaputra	

1 Choose the correct words from the box to fill in the gaps. (Hint: the map will help you.)

Bangladesh is in South-east _____. It is surrounded by

_____ and is just to the south-east of _____. There are

mountains in Nepal, called the _____. Two big rivers have deltas in

Bangladesh. They are called the _____ and the _____.

The capital city of Bangladesh is _____. The part of the Indian Ocean to the

south is called the _____.

2 Bangladesh is an LEDC (**L**ess **E**conomically **D**eveloped **C**ountry) which makes it more difficult for them to deal with an emergency like flooding. Explain why you think this is.

_____

_____

3 Imagine that you are involved in a flood in a country like Bangladesh. How might the help you get be different to the help given in Tewkesbury? Give reasons for your answer.

_____

_____

_____

_____

7.6 Protecting ourselves from floods

This is about ways to prevent floods.

1 There are several ways we can try to prevent floods happening. Some are shown below.

  a Shade in the boxes with pictures, each in a different colour.

  b For each picture, find the matching description and shade it in the same colour.

  c You should be left with reasons *why* flooding is reduced. Match these to the correct picture and description. Shade them the correct colour.

Build embankments or flood barriers around built-up areas.	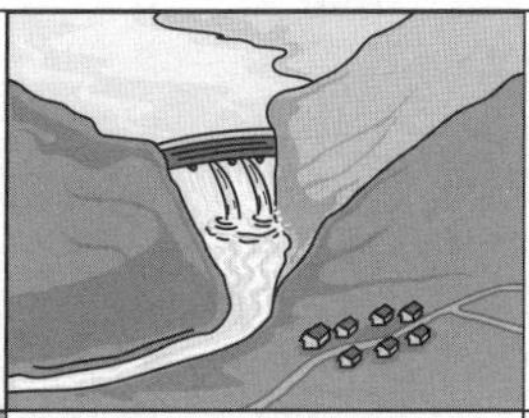	Build dams.	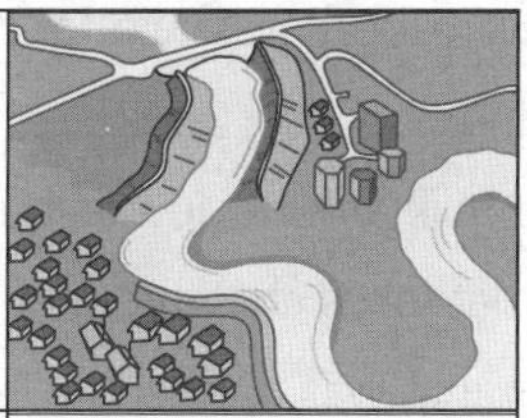
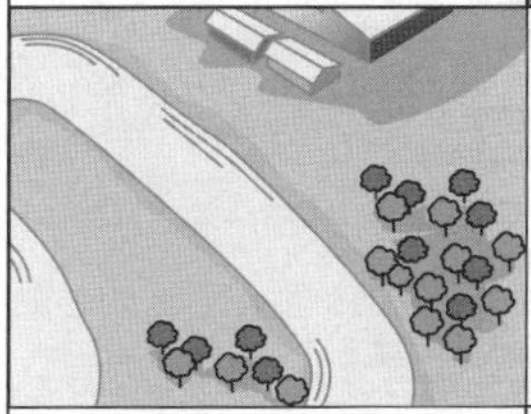	Make the river channel bigger.	These trap and store water, which can then be let out in a controlled way.	Plant more trees on the flood plain.
This speeds up the flow of the water, so there is less chance of a flood.	These keep the water in the river, and away from built-up areas.	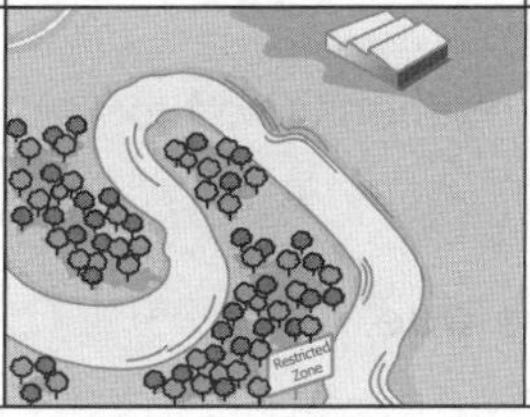	This will reduce surface runoff and increase interception.

2 There are other ways of trying to prevent flooding. Explain how each of these works.

  a improving street drainage

  _____

  _____

  b building pumping stations with storage basins

  _____

  _____

  c persuading farmers to let their fields along the river flood

  _____

  _____

7 And to finish ...

So, what have you learnt in this chapter? Let's see!

1 These statements are either **true** or **false**. Put a **tick** in the boxes for the true ones and a **cross** for the false ones.

	True	False
Floods happen when there isn't enough water in a place.	☐	☐
'Impermeable' means that water can't soak through.	☐	☐
'Saturated' means that there is more space for water to get in.	☐	☐
Cutting down trees makes floods more likely.	☐	☐
There were no floods in Tewkesbury in 2007.	☐	☐
The main reason for flooding in Tewkesbury was rain for 3 months.	☐	☐
Only Tewkesbury was affected by the floods in 2007.	☐	☐
The army helped to deliver bottled milk around Tewkesbury.	☐	☐
Gold Command is the police HQ that takes charge during an emergency.	☐	☐
COBRA is the name of a restaurant that was flooded in Tewkesbury.	☐	☐
In the UK, the Environment Agency builds flood barriers, checks water levels, and issues flood warnings.	☐	☐
Tewkesbury is at the confluence of two rivers, the Thames and the Wye.	☐	☐
Planting more trees in the drainage basin will help to reduce flooding.	☐	☐
Dams don't hold back water.	☐	☐
Straightening a river, and making it wider and deeper, are methods used to control flooding.	☐	☐

2 Now for all the **false** ones, re-write them so they become **true**.

_____

_____

_____

_____

_____

_____

_____

_____

_____

8.1 Geography and sport

This is about some of the links between geography and sport.

1 Some sports need stadiums or other buildings, others rely more on the natural environment. Look at the list of sports below. Put a tick in the box that you think is correct for each sport.

Sport	Needs a stadium/building	Needs natural environment
Fishing		
Snooker		
Mountain biking		
Speedway		
Golf		
Ice skating		
Judo		
Scuba diving		
Wrestling		
Climbing		
Ten-pin bowling		
Surfing		

2 What sport is important near where you live? Describe the sport and where it is played.

Name of sport _____

_____

_____

_____

3 Apart from the people who play your chosen sport, can you think of any other people who benefit may from it? Give reasons for your answer.

_____

_____

_____

_____

_____

8.2 The football business

This is about how the big football clubs make money – and spend it!

Does buying good players get clubs further up the Premiership?
Here are the Premiership placings in November 2005, together with the total amount spent on players' wages. Sit down before you read the Chelsea wage bill!

Club	Points on 7 Nov, 2005	Rank of points A	Annual wage bill	Rank of prices B	Difference between ranks C
Chelsea	31		£115 000 000		
Manchester United	21		£79 500 000		
Arsenal	20	3=	£69 900 000		
Tottenham Hotspur	20		£23 900 000		
West Ham United	18		£23 200 000		
Liverpool	16		£35 000 000		
Fulham	12	7	£25 000 000		

a Fill in column A by ranking the clubs by the number of points they have (two have been done for you).

b Complete column B in the same way.

c Now fill in column C by working out the difference between the two ranks. (Don't worry about plus or minus signs.)

The bigger the numbers in C, the bigger the difference between the clubs' position in the Premiership and the money they spent on players.

d Cross out the wrong words in these sentences.

The points ranking of a club is sometimes/often/always the same as the rank of a club by its wages bill.

Tottenham has/doesn't have a higher points ranking than its ranking for wages.

Liverpool should/shouldn't be doing better in the Premiership if you look at how much it's spending.

West Ham is/isn't getting value for money from its players.

e Based on what you've worked out so far, answer this question and explain your answer.

'If you were the chairman of a football club which was very short of money, would you sell a £30 million player to solve your money problems?'

_____

_____

_____

8.3 ▸ Liverpool FC is moving home

This is about Liverpool's move to its new stadium – and some of the impacts of the move.

1   When choosing a location for a football stadium, there are lots of special requirements that the club needs to think about. All these things together are called 'location requirements'.

   Brainstorm the things you'd need to look for if you were choosing a site for a Premier League club. (One has been done for you.)

> Will the public transport be able to cope?

2   The Stanley Park redevelopment has many impacts, on both Liverpool FC and the local residents.

   a   First colour in these top four speech bubbles, each a different colour.

> On the one hand, the move is a good thing for the club because …

> On the other hand, the move is a bad thing for the club because …

> On the one hand, the move is a good thing for the local community because …

> On the other hand, the move is a bad thing for the local community because …

   b   Now look at the speech bubbles below. Two of them discuss positive impacts of the move for the club. Colour them the same as the first speech bubble above.

   c   In the same way, colour the remaining statements.

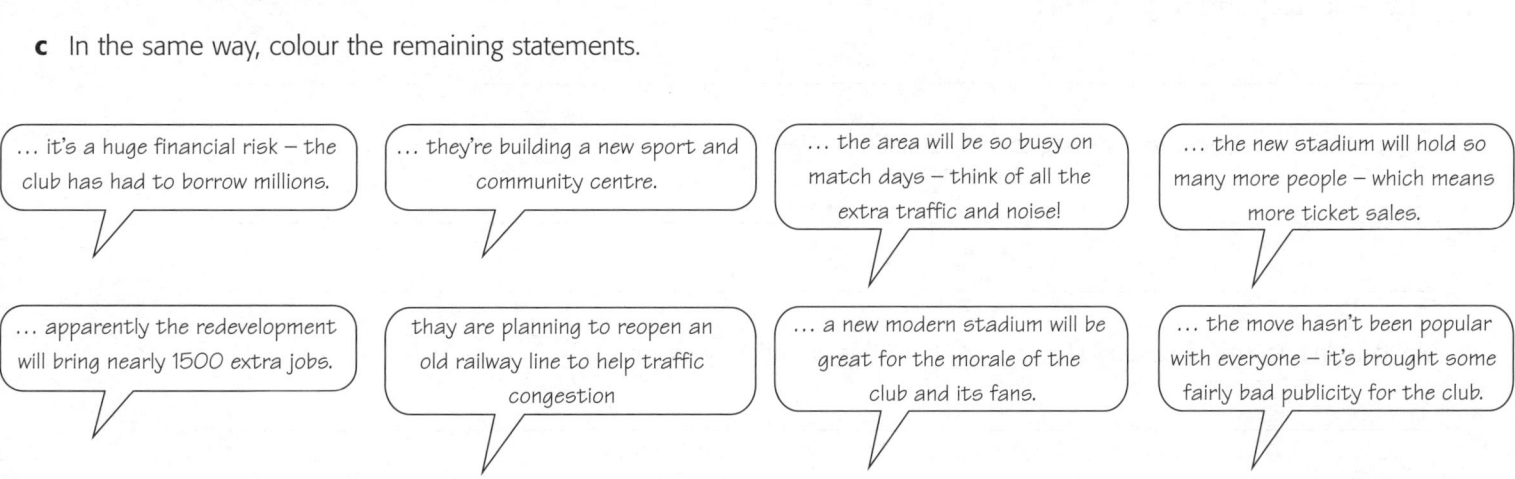

> … it's a huge financial risk – the club has had to borrow millions.

> … they're building a new sport and community centre.

> … the area will be so busy on match days – think of all the extra traffic and noise!

> … the new stadium will hold so many more people – which means more ticket sales.

> … apparently the redevelopment will bring nearly 1500 extra jobs.

> thay are planning to reopen an old railway line to help traffic congestion

> … a new modern stadium will be great for the morale of the club and its fans.

> … the move hasn't been popular with everyone – it's brought some fairly bad publicity for the club.

8.4 Who are the losers?

This is about how some 'football' workers are paid very unfairly.

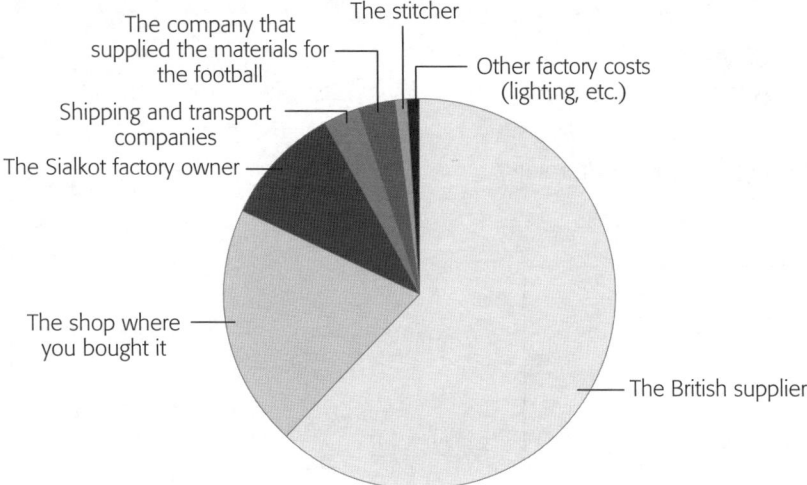

The company that supplied the materials for the football

The stitcher

Other factory costs (lighting, etc.)

Shipping and transport companies

The Sialkot factory owner

The shop where you bought it

The British supplier

1 This pie chart shows where the £50 spent on a football, made in Pakistan, is shared out.

a Where in the world are the businesses that take the two biggest 'slices'?

_____

b Where in the world are the businesses that take the two smallest 'slices'?

_____

c Why do people like Omar, who stitches footballs for a living, put up with being paid so little?

_____

2 Imagine that you stitch footballs for a living. You make four footballs a day, and earn £2.60 for your work. Your parents can't work, your father is ill and your mother has to look after your younger brother and sister.

a What health problems might you get?

_____

b How do you think you'd feel, having to sew footballs all day?

_____

_____

c Why couldn't you go to school? How do you think this will affect your future?

_____

_____

8.5 ▷ London 2012

This is about how the impact of the Olympic Games will be felt in other places as well as London.

1 Some of the Olympic events will be held away from London. The map shows where some of these are taking place. Choose the correct label from the box to complete the map.

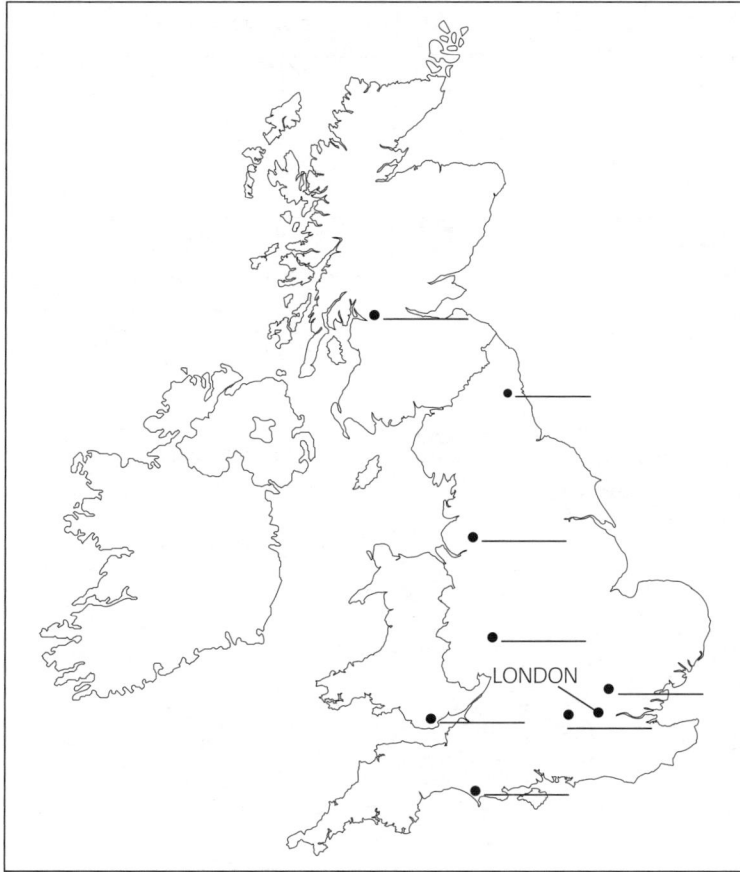

Hampden Park, Glasgow
Dorney Lake, Windsor
Millennium Stadium, Cardiff
Old Trafford, Manchester
Weald Country Park, Essex
St James' Park, Newcastle
Villa Park, Birmingham
Weymouth and Portland

LONDON

2 Weymouth and Portland will hold the sailing events. Give two good points and two bad points about living and working there when the Olympics are taking place.

Good points

_____

_____

_____

Bad points

_____

_____

_____

8.6 Making the Olympics sustainable

This is about how the organisers of the Olympic Games are trying to make each one more sustainable than the last.

A In Barcelona in 1992, the seafront area was totally changed and now offers leisure opportunities for visitors and residents.

B In Sydney in 2000, the Olympic Park was built on an old industrial area full of toxic waste.

C The "dressing up" of the city for the Games leaves other areas forgotten.

D London 2012 will use energy efficient buildings that will cut greenhouse gases.

E The Games only last two weeks – more temporary buildings could be used.

F A lot of solar energy was used at the Sydney Olympic Village.

G Beijing 2008 is trying to teach people to be more aware of the environment.

H Could the money not be better spent on health or education?

I London aims to offer a zero-waste, low-carbon Games.

J The Chinese have budgeted $23 billion for the 2008 Olympic Games – seven times more than the Sydney Games, and 32 times what Los Angeles spent for the 1984 Games!

1 a Colour in three statements that you think will best lead to a more sustainable Olympic Games.

b Now choose one of those reasons. Fill in this writing frame to explain why you think it will lead to a more sustainable Games.

Statement _____ will lead to a more sustainable Olympic Games.

This is because: _____

_____

_____

c I think that making the Games sustainable is important because:

_____

_____

_____

8 ◄ And to finish ...

Time to find out how much you have learnt about the business of sport around the world.

1 Match the figures in the boxes to the statements below.

| 65p | 80 million | 60 000 | £15 000 | 1892 |

| 40 million | 17 000 | 9 000 | £9 million |

What Omar gets for each ball he makes

The height of the new Liverpool stadium

The capacity of the new Liverpool stadium

What a top footballer earns in a week

The number of footballs made each year in Pakistan

The number of athletes and officials taking part in the 2012 Olympics

The number of workers employed on building the Olympic village

The overall costs of the 2012 Olympics

When Liverpool F.C. was founded

2 Do you think it is unfair that the people who make footballs should earn so little while the top football players earn so much?

_____

_____

_____

_____

_____

_____

9.1 ◀ A slice through the Earth

This is about the three layers that make up the Earth.

1 This diagram shows the layers that make up the Earth.

   a Choose the correct words from the box to label the diagram.

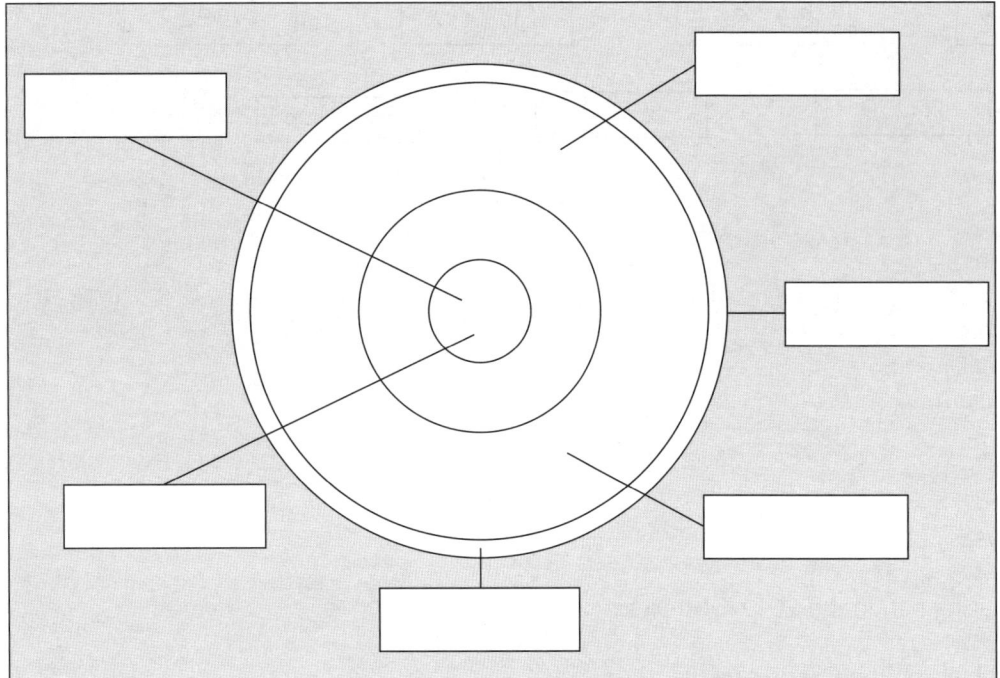

solid	the mantle	liquid
the mangle	the crust	a thin solid layer
the core	the gore	the edge-bit

   b Now colour the core yellow, the mantle orange, and the crust brown.

Finished? Try these …

2 Why are there convection currents in the mantle?

_____

_____

_____

3 There are two types of crust, continental and oceanic crust. What are the differences between them?

_____

_____

9.2 ◀ **Our cracked Earth**

**This is about where the Earth's plates are – and their link with earthquakes
and volcanoes.**

Why do earthquakes and
volcanoes happen mostly
in the same place?

1 Write a reply to this question. Use these words to help you.

plates	crust	margins
moving	convection	

Well, it's because ...

_____

_____

_____

_____

_____

_____

_____

2 Why don't earthquakes and volcanoes happen in some places?

_____

_____

3 A quick quiz! Cross out the wrong words, fill in the gaps or unscramble the jumbled words.

a The Earth's surface is broken into large pieces, like a cracked eggshell; the pieces

are called _____.

b The names of two large plates are c f i i c a p _____ and

s n i a a r e u _____.

c The plate I live on is called _____.

d Plates all move in the same direction / move in different directions.

9.3 How are the plates moving?

This is about how the Earth's plates are moving – and how their movements produce earthquakes, volcanoes, and even mountains!

1 These pictures show the different kinds of plate margins.

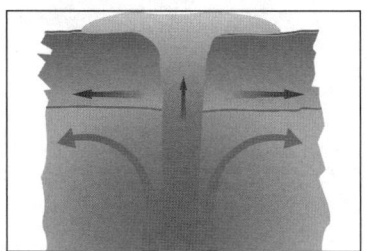 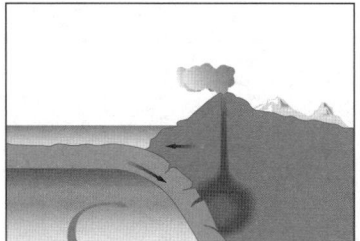 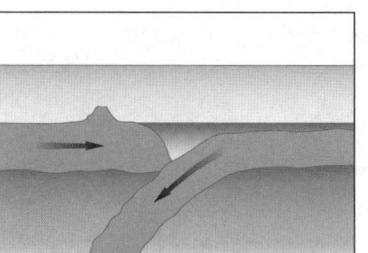 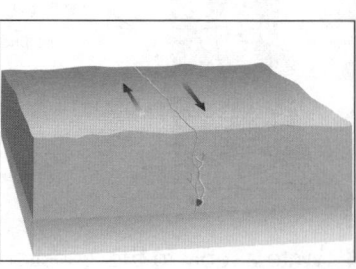

_____   _____   _____   _____

_____   _____   _____   _____

_____   _____   _____   _____

_____   _____   _____   _____

  a Label each diagram. Choose from:
- plates are moving apart
- plates are sliding past each other
- one oceanic plate is going under a continental one
- two continental plates pushing into each other

  b Draw a red star on the diagrams where you would expect earthquakes to happen.

  c Label any volcanoes you would expect.

  d Label the fold mountains.

  e Label the new crust that's being formed.

2 Which diagram do you think shows a **constructive** (building) margin? Why?

_____

_____

3 Which diagram do you think shows a **destructive** (destroying) margin? Why?

_____

_____

9.4 ◄ **Earthquakes**

This is about what earthquakes are, and how they are measured, and what damage they do.

Answer these questions to find the words that are hidden in the wordsearch.

1 Waves that are made by earthquakes _____

2 Small earthquakes after the main one _____

3 The point inside the crust where the earthquake started _____

4 The point on the surface right above where the earthquake started

_____

5 Amount of energy that an earthquake gives out _____

6 A machine that measures the magnitude of the earthquake _____

7 The scale that tells us how strong an earthquake is _____

8 A tidal wave caused by earthquakes _____

9 What an earthquake makes the ground do _____

10 If gas pipes fracture there might be _____

r	v	e	e	e	o	u	t	z	w	h	p	u	w	c
w	e	f	m	f	a	c	s	u	s	a	x	f	g	j
w	p	t	o	v	x	p	u	m	y	e	m	j	w	a
v	v	c	e	h	s	c	n	i	d	g	r	q	p	e
r	u	z	b	m	f	f	a	m	s	d	e	i	d	x
s	i	b	a	b	o	r	m	k	m	r	r	u	f	h
q	v	c	t	c	g	m	i	y	p	y	t	d	n	b
v	h	u	h	f	x	t	s	w	u	i	n	b	e	m
o	d	a	y	t	d	q	c	i	n	o	e	s	e	n
e	k	a	h	s	e	v	s	g	e	b	c	k	t	m
h	a	e	t	y	o	r	a	q	p	s	i	m	l	r
t	p	s	e	i	s	m	i	c	p	w	p	k	f	e
s	k	c	o	h	s	r	e	t	f	a	e	c	x	e
d	x	y	f	x	c	x	v	n	d	p	s	n	k	u
b	u	z	p	p	v	h	o	h	c	l	j	y	y	r

9.5 ▸ Earthquake in Pakistan

This is about the damage caused in the Pakistan earthquake in 2005.

The earthquake factfile

date	Saturday 8 October 2005
time	8.50 am
magnitude	7.6 on the Richter scale
epicentre	inside Pakistan-controlled Kashmir, about 20 km from Balakot
damage	over 74 000 people killed, over 106 000 injured, and over 3.3 million people left homeless, in Kashmir and northern Pakistan
financial cost	nearly £3 billion

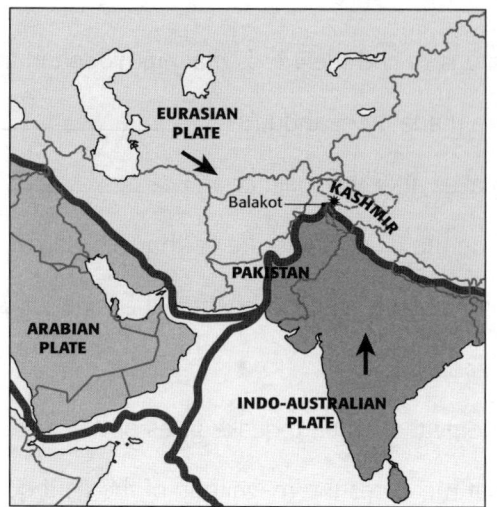

1 Using the information from the map and the Factfile, add words to the speech bubble to complete what the two people are saying.

This is what caused the earthquake.
_____
_____
_____
_____

The damage was terrible.
_____
_____
_____
_____

2 Look at the table below. Put a tick in the column to show if what happened could have been stopped or not.

Event	Could not be stopped	Could be stopped
1 The two plates are colliding and causing a lot of strain		
2 On 8 October the strain got too much		
3 At 8.50 in the morning the ground shook		
4 As the house collapsed behind them they could see people crying		
5 At the collapsed school people were trying to move the rubble with their hands		
6 Landslides and poor roads made it hard for help to get through		
7 Survivors waited for days with little food or shelter		
8 Without doctors many died from their injuries		
9 Over 74 000 people were killed		

9.6 Tsunami!

This is about the damage tsunamis do to the environment, the economy and people.

 A
 B
 C
 D

1 Choose one of the photographs and describe the damage that you can see. Write the letter of the photograph in the space.
   Photograph [   ]

   _____

   _____

   _____

   _____

2 Complete the table below to describe the damage caused to the environment, the economy and people.

	Natural (environment)	Economic (business/money)	Social (people)
Photograph A			
Photograph B			
Photograph C			
Photograph D			

9.7 ◀ **Volcanoes**

This is about what volcanoes are, and the damage an eruption can do.

Volcanoes happen on the edges of plates, where one plate goes under another.
Sometimes the plate suddenly starts moving and you get a new volcano where there
wasn't one before.

This is what happened in Mexico …

Draw a picture for each of the boxes in this cartoon strip.

1 It was 1943. In a flat maize
(corn on the cob) field in
Mexico…

2 … smoke and ash were
seen coming from the
ground.

3 The ground cracked and
there was a terrible smell
like rotten eggs.

4 By the next morning the
cone of ash was as tall as
a house.

5 Over the next year the cone
built up with layers of ash
and lava to make a new
volcano. It was named after
the nearest village – Paricutin.

6 The mountain erupted for
9 years! By the end it had
covered 10 square miles
in lava and killed over
1000 people.

7 The lava and ash caused
damage during the eruption
but left the soil fertile (good
and rich) so now lots of crops
can be grown and livestock
(farm animals) graze the land.

8 Now Paricutin is nearly
400 metres high

9.8 ◀ Montserrat: living with an active volcano

This is about how an erupting volcano has changed a Caribbean island forever.

Before the volcano on Montserrat started erupting in 1995 people thought it was extinct. But it wasn't. It was only dormant. That means 'asleep'.

In July 1995 the volcano began to waken – the dust and ash were our warning. Scientists moved in and started monitoring the gases coming out and the changes in the shape of the mountains. They also had machines called seismographs to measure sany tiny vibrations underground.
In August 1995, we started to evacuate the south of the island. We had to live in halls and churches, sharing toilets… I'm not sure we'll ever be able to return to the south because the volcano is still active. There was a bad eruption in July 2001 and another in July 2003.

In April 1996, everyone was told to leave Plymouth, our capital city. Then, in June 1997, the south of the island was covered by rivers of hot ash, gases, mud and rock. Scientists call these pyroclastic flows. The fires were huge! 19 people were killed, because they'd refused to stay away from their homes. We've lost two-thirds of our houses and three-quarters of our roads. More than half of the island's people are leaving. The island is losing so much money because there aren't any tourists any more. There used to be about 2000 a year.

1 Use these speech bubbles to fill in the time-line for Montserrat.

- 1995
- 1996
- 1997
- 1998
- 1999
- 2000
- 2001
- 2002
- 2003

2 What have the effects of the eruptions been on:

a jobs?

_____

b people's homes and other buildings?

_____

9.9 Coping with earthquakes and eruptions

This is about how we cope with earthquakes and volcanoes – and why some countries find it harder than others.

1 Cross out the wrong word in this sentence.

The wealthier/poorer a country, the more effective it's likely to be in responding to disasters.

2 These are all things that were done after the eruptions started on Montserrat. Underline the **short-term responses** in *red* and the **long-term responses** in *green*.

- Vulcanologists (volcano scientists) monitored the volcano, and made plans to move people to safety.

- People have moved back to the island. By 2005 the population was over 8000.

- The Red Cross has built a new home for the elderly.

- The UK Government sent £17 million in emergency aid, including temporary buildings, water purification systems, and expert people too.

- In 2005, the south of the island remains out of bounds and vulcanologists are still monitoring the volcano.

- The UK Government has funded a 3-year redevelopment programme for houses, schools, medical services, infrastructure and agriculture. People have also been offered mortgages so they can start new businesses. It cost £122.8 million.

- Charities like the Red Cross set up temporary schools.

- People are evacuated to the north, and some to other countries. By November 1997, the island's population has fallen to 3500.

- The population structure has changed because many young people have made new lives elsewhere, whilst many older people either never left, or have moved back to the island.

- People are evacuated on boats paid for by the UK and the USA, as well as using their own boats. The British navy also evacuate some people.

- Tourists may come back and the volcano itself may become a tourist attraction.

- The USA sends in troops to help with the evacuation.

- Some vegetation is starting to re-grow in the south of the island. The soil will eventually become fertile as the ash and lava break down.

- Charities send emergency food for farm animals.

9 ⟩ **And to finish ...**

Time to find out how much you've learnt about plates, earthquakes and volcanoes!

1 The Earth's crust is broken into pieces called _____

2 Things that are done straight after a natural disaster are _____

3 Soil that comes from weathered lava is _____

4 The solid centre of the earth is called the _____

5 Things that are done a while after a natural disaster are _____

6 Where the lava comes out. _____

7 The part of the island of Montserrat that was destroyed. _____

8 Magma is called _____ when it gets above the ground. _____

9 People rely on this to make sure they aren't hurt in an earthquake. _____

10 The name of the currents in the mantle that move the plates. _____

11 The proper name for a volcano scientist. _____

12 The point on the ground straight above the focus. _____

13 The scale that's used for measuring the size of earthquakes. _____

14 A rapid flow of dust, ash and gases is called a _____ flow.

15 The two plates meeting at Montserrat are the Caribbean and the _____

16 The thin outer layer of the Earth. _____

17 Balakot (where there was a huge earthquake) is in _____.

18 The point where an earthquake happens, underground. _____

19 Where the magma is stored in a volcano. _____

20 Small quakes after the main one. _____

21 A series of waves, set off by an earthquake in the ocean floor. _____

warning	magma chamber	North American	Iran	aftershocks	epicentre
focus	vent	short-term responses	crust	vulcanologist	pyroclastic
tsunami	core	plates	long-term responses	south	fertile

OXFORD
UNIVERSITY PRESS

Great Clarendon Street, Oxford OX2 6DP

Oxford University Press is a department of the University of Oxford. It furthers the University's objective of excellence in research, scholarship, and education by publishing worldwide in

Oxford   New York

Auckland   Cape Town   Dar es Salaam   Hong Kong   Karachi   Kuala Lumpur
Madrid   Melbourne   Mexico City   Nairobi
New Delhi   Shanghai   Taipei   Toronto

With offices in

Argentina   Austria   Brazil   Chile   Czech Republic   France   Greece   Guatemala
Hungary   Italy   Japan   South Korea   Poland   Portugal   Singapore   Switzerland
Thailand   Turkey   Ukraine   Vietnam

Oxford is a registered trade mark of Oxford University Press
in the UK and in certain other countries

Authors: Anna King, Jack Mayhew, Susan Mayhew, Justin Woolliscroft

The moral rights of the author have been asserted

Database right Oxford University Press (maker)

First published 2006
New edition published 2008

British Library Cataloguing in Publication Data

Data available

ISBN 978 0 19 913502 8

20 19 18

Printed by Ashford Colour Press Ltd., Gosport

Acknowledgements

The publisher and authors would like to thank the following for permission to use photographs and other copyright material:

P4 (left) Julien Grondin/ Shutterstock; P4 (middle left) Merten Merten/ Mauritius Die Bildagentur Gmbh/ Photolibrary; P4 (middle right) Douglas Sterling/ Arcangel Images Ltd/ Photolibrary; P4 (right) Ulrike Hammerich/ Shutterstock; P6 Jeremy Horner/ Alamy; P11 Bob Croxford/ Corbis; P25 Chris Gascoigne/ View Pictures/ photolibrary; P28 Jason Hawkes/ Corbis; P53 (top) Richard Wainscoat/ Alamy; P53 (bottom) Garry L Tong/ Corbis; P57 Barry Bland/ Alamy; P65 (left) AFP/ Getty Images; P65 (middle left) AFP/ Getty Images; P65 (middle right) Associated Press; P65 (Right) Associated Press.

The Ordnance Survey map extract on page 15 is reproduced with the permission of the Controller of Her Majesty's Stationery Office © Crown Copyright.

Every effort has been made to contact copyright holders of material reproduced in this book. Any omissions will be rectified in subsequent printings if notice is given to the publisher.

www.OxfordSecondary.co.uk

Orders and enquiries to Customer Services:
tel. 01536 452620 fax 01865 313472

geog.1 workbook

3rd edition

This workbook is ideal for
homework and independent
study. It provides support for
every double-page spread in
the geog.1 students' book.

geog.123 is a three-book course
for the National Curriculum at
Key Stage 3.

Did you know?
♦ You are living
on a moving
slab of rock.

What if ...
♦ ... you
owned a river?

OXFORD
UNIVERSITY PRESS

Orders and enquiries to Customer Services:
tel. 01536 452620 fax 01865 313472
schools.enquiries.uk@oup.com

ISBN 978-0-19-913502-8

9 780199 135028

www.OxfordSecondary.co.uk